CO-CITIES

Urban and Industrial Environments

Series editor: Robert Gottlieb, Henry R. Luce Professor of Urban and Environmental Policy, Occidental College

For a complete list of books published in this series, please see the back of the book.

CO-CITIES

**INNOVATIVE TRANSITIONS TOWARD JUST
AND SELF-SUSTAINING COMMUNITIES**

SHEILA R. FOSTER AND CHRISTIAN IAIONE

THE MIT PRESS   CAMBRIDGE, MASSACHUSETTS   LONDON, ENGLAND

© 2022 Massachusetts Institute of Technology

This work is subject to a Creative Commons CC-BY-ND license.
Subject to such license, all rights are reserved.

The open access edition of this book was made possible by generous funding from the MIT Libraries.

The MIT Press would like to thank the anonymous peer reviewers who provided comments on drafts of this book. The generous work of academic experts is essential for establishing the authority and quality of our publications. We acknowledge with gratitude the contributions of these otherwise uncredited readers.

This book was set in Stone Serif and Avenir by Westchester Publishing Services.
Printed and bound in the United States of America.

Library of Congress Cataloging-in-Publication Data

Names: Foster, Sheila R., 1963– author. | Iaione, Christian, author.
Title: Co-cities : innovative transitions toward just and self-sustaining
   communities / Sheila R. Foster and Christian Iaione.
Description: Cambridge, Massachusetts : The MIT Press, [2022] | Series: Urban
   and industrial environments | Includes bibliographical references and index.
Identifiers: LCCN 2021062248 (print) | LCCN 2021062249 (ebook) |
   ISBN 9780262539982 (paperback) | ISBN 9780262361910 (epub) |
   ISBN 9780262369930 (pdf)
Subjects: LCSH: City planning. | Civic improvement. | Commons. | Municipal
   government. | Infrastructure (Economics) | Public-private sector cooperation.
Classification: LCC HT166 .F675 2022 (print) | LCC HT166 (ebook) |
   DDC 307.1/216—dc23/eng/20220519
LC record available at https://lccn.loc.gov/2021062248
LC ebook record available at https://lccn.loc.gov/2021062249

CONTENTS

ACKNOWLEDGMENTS ix

INTRODUCTION 1
1 RETHINKING THE CITY 33
2 THE URBAN COMMONS 61
3 THE CITY AS A COMMONS 103
4 URBAN CO-GOVERNANCE 149
5 THE CO-CITY DESIGN PRINCIPLES 191
   CONCLUSION: NEW CO-CITY HORIZONS AND CHALLENGES 219

APPENDIX 239
NOTES 255
REFERENCES 257
INDEX 293

ACKNOWLEDGMENTS

This book is the result of a wide collaborative effort. Although based in our individual and joint scholarship, this book benefited from our close collaboration with a number of academics, students, networks, and dedicated graduate and postgraduate fellows. The Co-Cities project started initially through collaboration with Marco Cammelli, Emeritus Professor of Administrative Law at the University of Bologna, former President of the Fondazione del Monte di Bologna e Ravenna, the Transformative Actions Interdisciplinary Laboratory (TrailLab) of the Catholic University of Milan, in particular Professor Ivana Pais and Michela Bolis, and the International Association for the Study of the Commons (IASC). It then drew on years of dialogue, exchange and learning with scholars, practitioners, and undergraduate and graduate students at Luiss University, Sapienza University, Fordham University, and Georgetown University.

We want to thank in particular Bill Treanor, the Dean of Georgetown Law, for providing indispensable research support. Special thanks are also due to Leonardo Morlino, Emeritus Professor of Political Science at Luiss University, for his tireless and patient guidance on the theoretical framework and the methodological approach reflected in the early part of this project and to Claudio Rossano, Raffaele Bifulco, Paola Chirulli, Fulvio Cortese, Giacinto Della Cananea, Annalisa Giusti, Bernardo Mattarella, Giuseppe Piperata, Aldo Sandulli, Maria Alessandra Sandulli, and Paolo

Stella Richter, for sharing their scholarly wisdom on public and administrative law.

Thank you to the over two hundred participants at the City as a Commons IASC Conference (Bologna 2015) and in particular the scholars and practitioners who shared their presentations and papers with us and helped us spread the unique body of knowledge on the urban commons generated by the conference that has shaped our thinking. We are specifically thankful for the contributions of the keynote scholars and researchers at the conference, including David Bollier, Tine de Moor, Paola Cannavò, Silke Helfrich, and Ezio Manzini. We are thankful to the city of Bologna and in particular to Virginio Merola, Matteo Lepore, Giacomo Capuzzimati, Luca Rizzo Nervo, Giovanni Ginocchini, Michele D'Alena, Ilaria Daolio, Roberta Franceschinelli, Anna Rita Iannucci, and Donato Di Memmo for their support in organizing the 2015 conference and the connected action-based research program and policy experimentation initiated in 2011 and developed until 2017 under the umbrella of the Co-Bologna program.

We launched the empirical Co-Cities project on the heels of that conference and the connected action-based research program and policy experimentation, with the help of Michel Bauwens and Vasilis Niaros (P2P Foundation Research Fellow) who contributed to the first phase of our case study surveys. For case studies in Latin America, we relied on the excellent suggestions offered to us by Thamy Pogrebinschi. We also relied upon the invaluable data and analysis collected by her and her research team on LATINNO, innovations for democracy in Latin America.

For case studies on sharing cities worldwide we drew upon some of the examples in the book *Shareable Sharing Cities: Activating the Urban Commons*. We are thankful to Shareable co-founder Neal Gorenflo for his support and for being a constant source of inspiration throughout the last years. We benefited tremendously from our mutual exchange with Daniela Patti and Levente Polyàk, co-founders of Eutropian, who nurtured our interest in sustainable urban regeneration processes. We have also learned a great deal from Eutropian's research projects and found particularly useful the case studies contained in the book *Funding the Cooperative City*. We shared the initial part of this research journey with Gregorio Arena, Daniele Donati, and Fabio Giglioni at Labsus.org.

ACKNOWLEDGMENTS

We are deeply grateful to Nicoletta Levi, Valeria Montanari, Lanfranco De Franco, and Luca Vecchi, leading policymakers in Reggio Emilia; Gianni Ferrero, Fabrizio Barbiero, Walter Cavallaro, Ilda Curti, Alice Zanasi, Tiziana Eliantonio, Laura Socci, Alessandra Quarta, Ugo Mattei, Antonio Vercellone, Rocco Albanese, Simone D'Antonio, and Paolo Testa, key players in the Co-City Turin UIA project; Nicola Masella, Fabio Pascapè, Alberto Lucarelli, Carmine Piscopo, Giuseppe Micciarelli, Roberta Nicchia, and Maria Francesca De Tullio, key figures in the Naples urban commons policymaking and advocacy who led also the Civic eState URBACT project (which helped us learn from the professional expertise of exemplary urban civil servants and activists such as Magdalena Skiba and Michal Zorena from Gdansk; Natalie van Loon from Amsterdam; Gala Pin, Laia Fornet, Albert Gomez Martin, and Mauro Castro from Barcelona; Emma Tytgadt, Yoko Gesels, and Eleke Langeraert from Gent; Marius Homocianu from Iasi; and Peter Fomela from Presov); Mauro Annunziato and Claudia Meloni, lead scientists at ENEA the Italian National Research Agency on Sustainable Development, who enabled our research on smart communities and energy communities; Luca Talluri, Alessandro Almadori, and Luca Montuori, who supported our research endeavor to reconceive public and social housing as commons; Massimo Alvisi and Gerlando Iorio, who involved us in the Co-Battipaglia project and made us think about the key role urban planning may play in enabling the urban commons; and Paola Verdinelli, Paola Cannavò, Dario Minghetti, and Urio Cini, who played a key role in the first river participatory foundation and the first neighborhood cooperative incubated in Rome through the Co-Roma .it platform and project. They all contributed to give us the opportunity to collaborate in a collective effort to put the theory of the urban commons into practice in stimulating testbeds like Bologna, Reggio Emilia, Turin, Naples, and Rome.

We express our gratitude as well to the Fordham University Urban Consortium and the Urban Law Center at Fordham University that welcomed Christian Iaione as a faculty fellow and welcomed Elena De Nictolis as a visiting PhD research fellow. Professor Nestor Davidson, current director of the Urban Consortium and the Urban Law Center, has been a consistently supportive interlocutor and we remain inspired by his work on urban law, the sharing economy in cities, and local administrative law. We also want to acknowledge Margie Mendell, co-founder of the Karl Polyani Institute of Political Economy at Concordia University, whose work on political economy and on economic democracy has been a critical intellectual companion in developing the Co-Cities theoretical framework. We are also grateful for the support and feedback received from Aaron Maniam and the team at Oxford Urbanists and to Pakhuis De Zwijger (Amsterdam) and the European Cultural Foundation (ECF) for involving us in inspiring academic and policy conversations on the future of cities and democracy in Europe.

We acknowledge and extend a special thanks to Alicia Bonner Ness, who masterfully coordinated and moderated an international workshop to validate the Co-City protocol at the Rockefeller-sponsored Bellagio Conference ("Accelerating Citywide Entrepreneurship: An Exercise in the Co-City Approach"). We are grateful to the Bellagio Conference Program staff for their hospitality and to the retreat participants for taking part in such a challenging experiment. Those participants included representatives from the city of Amsterdam, Netherlands (Urban Innovation Office); the city of Barcelona, Spain (Regidoria de Participaciò i territorio); the city of Boulder, Colorado (Chief Resilience Officer [CRO]); the city of Turin, Italy (the Co-City project); the city of Madison, Wisconsin; the city of New York (NYCx Co-Labs program of the Mayor's office); Habitat International Coalition; the National Association of Italian Cities (ANCI); Cooperation Jackson (Jackson, Mississippi); Archiafrika (Accra Ghana); the German Marshall Fund of the United States (the Urban and Regional Policy Unit); the Brookings Institution (the Project on 21st Century City Governance); the Laboratory for the City, Laboratorio para la Ciudad (Mexico City); and SPUR (formerly known as the San Francisco Bay Area Planning and Urban Research Association).

The contribution of LabGov.City research associates and research fellows for building the database and carrying out the empirical analysis was indispensable: Elena De Nictolis, without whom the Co-Cities project or this book would have been possible, coordinated the methodology and data mining and coding team (2015–2021); Chiara De Angelis provided support as lead research associate and designed the digital tool that hosts the Commoning.city map and the Co-Cities data set (2015/2018); Cosima Malandrino supported the data analysis and communication strategy of the report, first as a graduate intern and later as research associate (2018–2020); and Alessandro Antonelli provided crucial support in the latest round of data collection and coding as well as curator of the graphic design and digital layout of the appendix for this book (2019–2021).

We owe a special debt of gratitude to many research fellows who nurtured this research directly or indirectly through conversations and correspondence, by providing comments and feedback on earlier drafts of this work, or by cooperating with LabGov.City research projects and conducting independent research on some of the case studies included in the Co-Cities dataset. Crucial research carried out at Georgetown University by doctoral fellow Chrystie Swiney and McCourt Public Policy School master students Sumedha Jalote, Zezhou Cai, and Shamila Kara included interviews with case study participants, data entry, data collection, and detailed case studies' analysis in the US, India, and China. A precious contribution came from Monica Bernardi, Fabiana Bettini, Chiara Prevete, and Anna Berti Suman who conducted with us careful and passionate research, both normative and empirical and also provided support with data collection and analysis of projects/public policies (in Seoul and Boston; Community Land Trusts in Europe and the UK; social and public housing in Italy; citizen science in EU cities, respectively).

The work of LabGov.city research associates during their internship and collaboration on the Co-Cities project was instrumental for the production of the dataset. Their support on surveying cases, conducting interviews, and coding data was invaluable, and we owe them a special debt of gratitude for their generosity in sharing knowledge and research skills as well as their proactivity: Lucia Paz Errandonea (2017), Lucille Reynaud (2019), Carlo Epifanio (2019), Anouk Jeanneau (2019–2020), Federica Muscaritoli (2021), Micol Sonnino (2021), and Riccardo Negrini (2021). We would like to express our deep appreciation to Sofia Croso Mazzuco and Benedetta Gillio for their contribution as research assistants to data entry and data collection over many years of the project.

We want to extend a special note of thanks to the TAs and students of the Urban Law and Policy course, the Smart Cities course, the Governance of Innovation and Sustainability course, and the Law and Policy of Innovation and Sustainability course within the MSc Law, Digital Innovation and Sustainability (Luiss University, Department of Law, and

Department of Political Science). Among the others, we would like to express our appreciation to Alessandro Piperno, Luna Kappler, Alberica Aquili, Azzurra Spirito, Erasmo Mormino, Pier Paolo Zitti, Tommaso Dumontel, Federica Muzi, Alessio Ciotti, Giannandrea Ingallinera, Margherita Frabetti, Sofia Brunelli, Clara Wrede, Mohtas Modier, Omolola Akiniyi, Benedicta Quarcoo, Albert Gimenez, Nathan Senise Volpe, Francesco Palmia, Gresia Bernardini Marino, Mattia Lupi, Paolo Marro, Serena Ragno, Giulia Balice, Federico Pieri, Elisa del Sordo, Martina Rotolo, Guglielmo Pilutti, Marina Gascòn, Marta Pietro Santi, Greta Bertolucci, Charlotte Poligone, Zita Kučerová, and Riccardo Negrini. Their support in research grounds as well as through the research papers, class discussions over the course of the academic years between 2017 and 2021 have been of help in a variety of ways.

We are deeply thankful to Luiss University Rector Andrea Prencipe; Vice-President Paola Severino; Director General Gianni Lo Storto; Deputy Rectors Simona Romani, Luca Giustiniano, Raffaele Marchetti, Antonio Gullo, Stefano Manzocchi, Fabiano Schivardi, and Luiss Research, PhD, Recruiting, and Partnerships offices; and Luiss Law School key figures Antonio Nuzzo, Antonio Punzi, Bernardo Mattarella, Antonio Gullo; and Luiss Law School staff for hosting at Luiss University several episodes of our intellectual journey and in particular the Urban Law Conference on the Theory and Pedagogy of Urban Law, hosted in cooperation with the Fordham Urban Law Center. Invaluable contributions and insights came in that occasion from the exchange with Paola Chirulli, Anel Du Plessis, Antonia Layard, Alexandra Flynn, Sofia Ranchordas, Jean-Bernard Auby, Yishai Blank, Daniel Bonilla, Bernardo Mattarella, Andreas Philippopoulos, Aristide Police, Shitong Qiao, Aldo Sandulli, and Rich Schragger. Inspiration for this book and resources for its base and applied research activities derived from the Horizon 2020 projects OpenHeritage.Eu, EUARENAS.EU, Engage.Eu R&I, and Engage.EU Erasmus+.

Finally, our efforts to bring the co-city protocol to US cities benefitted immensely from a number of academic partners, local officials, civic organizations, and financial sponsors. Fordham Professor Oliver Sylvain, whose work on network equality and broadband localism has been influential for us, worked with us to develop the community-led broadband project in Harlem, New York City, using the co-city framework as described in this book. That project, funded by the National Science

Foundation, is a collaboration between Clayton Banks and Bruce Lincolon of Silicon Harlem, a community-based technology and innovation center, and researchers representing a number of universities: Olivier Sylvain from Fordham Law School; Rider Foley, Ron Williams, and the late Malathi Veeraraghavan from the University of Virginia; and Dan Kilper and Bryan Carter from the University of Arizona.

We also thank Professor Chris Tyson who served as CEO of Build Baton Rouge (formerly East Baton Rouge Redevelopment Authority) from 2017 to 2021 and who invited us to implement the co-city protocol as part of his effort to revitalize a four-mile segment of the Plank Road corridor of Baton Rouge, Louisiana. The Co-City Baton Rouge project, which continues to this day and is described in the book, was made possible at its inception with the support of New York University's Marron Institute for Urban Management. We are deeply thankful for the support of Professor Clayton Gillette, who served as the Director of the Marron Institute, whose work in the field of local government law and finance has been an inspiration to us. We also thank Manohar Patole, the Co-City Baton Rouge fellow and project manager, for the dedication, energy, creativity he has brought to implementing the model with our local and community-based partners.

This book is also an intellectual bridge with our colleagues at the Center for Civil Society and Governance at the University of Hong Kong, especially Winnie Law, Joyce Chow, and Wai-Fung Lam, and at the Department of Architecture of the Universidad Latina de Costa Rica, especially Margherita Valle. By embarking on this intellectual journey with us and creating respectively LabGov Hong Kong and LabGov Costa Rica, they opened the doors for challenging, adapting, and revising our framework to different contextual factors than those from which the Co-City approach originated. We look forward to working with them and with others, as this book really is the first step result of a wide collaborative effort.

This book is the result of a long and mutually enriching intellectual journey between two colleagues, engaged scholars, and friends. While a collaborative effort has led to the joint production of the research results presented in this book, if attribution is required the introduction, chapters 1 and 2 should be attributed to Sheila Foster, while chapters 3, 4, and 5 should be attributed to Christian Iaione. The conclusion should be jointly attributed to both authors.

INTRODUCTION

It is now a truism that we are an urbanized world. Today a majority (55 percent) of the world's people live in cities and their surrounding metropolitan areas, a trend that has been accelerating since the mid-to-late twentieth century (UN DESA 2019). It is estimated that by 2050 almost 70 percent of the world will live in cities and metropolitan areas. Cities are not only where most of the world's population will continue to physically concentrate but are also key centers of economic production. Consider that in the US, the ten most productive metropolitan areas alone contribute about 40 percent of the gross domestic product (GDP), and the top twenty metropolitan areas contribute over 50 percent of the US GDP (Perry 2018). This includes metropolitan areas and surrounding major cities like New York, Los Angeles, San Francisco, Boston, and Chicago. The New York City metro area alone constitutes about 10 percent of the US total GDP and produces 4 percent more economic output than the entire country of Canada (Perry 2018). The Los Angeles metro area, the second largest US metropolitan economy, produced just slightly less economic output than Mexico. Chicago's economy is the third largest metro economy in the US; its GDP is slightly higher than Switzerland's GDP.

Outside of the US, the numbers are as revealing. Cities such as Toronto, Mexico City, Tokyo, London, Paris, Stockholm, Tel Aviv, and Seoul generate anywhere from 18 percent (Toronto) to over 50 percent (Tel Aviv) of national GDP (Florida 2017a). According to some analyses, people living in large cities will account for as much as 81 percent of global consumption by 2030 and 91 percent of global consumption growth by 2015 to 2030 (Dobbs et al. 2016).

As the most highly educated and affluent populations agglomerate in cities, the working class and low-income workers and their families are increasingly being pushed out of the urban core and further away from the economic opportunities concentrated there (Ehrenhalt 2012). The spatial mismatch between where jobs are located and where the lowest income workers live, often on the periphery, is immense in some cities and metropolitan areas. America's inner-ring suburbs, for example, once a marker of suburban prosperity, are increasingly becoming home to poorer populations and immigrants as core cities become less affordable (Ehrenhalt 2012). Likewise, rural to urban migration into cities in the Global South fuels the expansion of informal occupations of peripheral urban areas and precarious urban settlements (Davis 2006).

As urban peripheries grow and expand outward, so too do stark wealth and resource disparities between neighborhoods within a city and between cities (and towns) within metropolitan regions. The result is high and worsening levels of income inequality, ethnic and economic segregation, increasingly unaffordable housing and food insecurity, and unequal access to amenities ranging from broadband networks to parks and green spaces.

While cities have become the heart of the growing problem of economic and spatial inequality, within them lie at least some of the solutions to this inequality and to other challenges that range from climate change to migration. Exactly how cities can meet these challenges is one of the core debates in contemporary urban literature. This literature offers several urban approaches or frameworks that capture how cities can function to make human life within them better, smarter, more efficient, more sustainable, and/or more equitable and inclusive.

One approach focuses on cities' need to compete for and attract concentrations of highly educated workers, highly skilled employees, and high tech and creative industries. Cities that attract the right combination of skill and talent and the industries that serve them are expected to reap and sustain high growth and economic prosperity. The positive spillover effects of these agglomeration economies also promise to create new classes of service workers and job opportunities for a broader class of residents.

Another approach focuses on technology as the main force shaping cities and making life better for the people who govern and live in them. In this vision, the city is a platform for the use of advanced technologies and data, equipping cities with improved features for a safe and more convenient urban existence. Technology can also more effectively enable urban residents to participate in local decision making and more efficiently deliver a range of goods and services. Equipped with sophisticated technological tools, cities would become digitally networked places, facilitating the sharing of urban infrastructure and digital platforms that more efficiently connect urban inhabitants to each other and to businesses and service providers.

One more prominent approach or framework views urban life through the lens of the right to the city and seeks to empower residents to collectively shape the city for its inhabitants. Cities, according to this framework, can be places that fully realize the right to adequate housing, universal access to safe and affordable drinking water and sanitation, equal access to quality education, and other public goods. Embedded in this approach is a city *for all*, where different populations have equitable and affordable access to basic physical and social infrastructure to foster prosperity and the sustainability of human settlements.

These three frameworks are the most prominent offered by urban scholars for understanding urban growth in cities today, ways to shape future cities, and the potential of cities to address some of our most pressing social and economic challenges. Each approach is rooted, at least in part, in reality. For example, we see evidence of urban agglomerations of knowledge and technology workers and the industries that depend on them in cities like San Francisco and Toronto; the emergence of digitally sophisticated *smart cities* in cities such as Amsterdam and New York City; and the codification of the *right to the city* in Brazil's City Statute, Mexico City's Right to the City Charter, and policies taking shape in Barcelona and other European cities.

In this book we offer another approach rooted in our decades-long investigation of over two hundred cities and direct involvement in crafting practices and policies in a few of them that enable public, community, civic, knowledge, and private actors to collectively create and then steward shared urban resources throughout the city. Our framework, the co-city, captures and reflects the ways that some cities are moving or being pushed toward embracing practices and policies that are fostering social innovation in urban services provision, spurring collaborative economies as a driver of local economic development, and promoting inclusive and equitable urban regeneration of blighted areas. We believe that the co-city framework, supported by an ongoing empirical project and the conceptual building blocks articulated in our scholarship over the last decade, is in part a challenge to existing urban vision frameworks and in part a refinement of them.

The co-city framework offers a new path forward—including new tools and practical approaches—to achieve some of the normative goals that animate, for instance, the right to the city and smart cities. At the same time, it is a response to the failures of the market-based agglomeration approach to cities that is associated with vast economic stratification and inequality in so many global cities.

As we write this, the COVID-19 pandemic and the movement against racial injustice have accelerated a rethinking of the urban landscape to meet the challenges of creating more resilient and just cities. New ideas and visions are emerging to meet these challenges. Ideas that range from the *15-minute city* that focuses the scale of urban planning to provide access to all human needs within a short walk or bike ride, to more equity-focused place-making approaches that reimagine public spaces to mitigate racial, ethnic, and gender divisions in cities. Although there is no one vision or panacea that can address all the challenges faced by contemporary urban life, the co-city framework offers something distinct in comparison with existing urban visions, as we set out in this chapter.

THE CITY AS A MARKET: URBAN AGGLOMERATION ECONOMIES AND CREATIVE CITIES

A prominent strand of urban research has drawn attention to the relationship between urban agglomeration, knowledge and creative workers, and urban economic growth in cities. Urban economists have long posited that the migration of workers with high levels of talent to amenities-rich locations is the dominant reason that cities and metropolitan regions grow and other regions remain stagnant or decline. Urban agglomeration economic theories trace to seminal works by Jane Jacobs (1969) and Edward Glaeser (1998, 1999), among others, to account for the fact that individuals move to cities not only to increase their wages but also to capitalize on the proximity to others like themselves.

The core claim of urban economists is that individuals seeking economic (and other) gains base their location decisions on where similarly *high human capital* individuals cluster. These agglomeration gains can include the ability to learn from other high knowledge workers and to acquire additional skills through information spillovers. Agglomeration economics suggests that individuals more *efficiently* acquire new skills in urban metropolitan areas because of the greater opportunities to interact with other highly educated and skilled people, thus increasing the rates of human capital accumulation, technological innovation, and ultimately, urban growth.

An important possible gain from agglomeration of high human capital individuals is *matching*, a form of labor market pooling in which workers have a greater likelihood of obtaining a good match between their skills and an employer, thereby increasing productivity and wages. For example, technology workers who cluster in regions with a concentration of technology firms will be better able to find firms that value their skills and talent. Similarly, firms will benefit when they are looking for specific skill sets or specialized labor force because they will have plenty to choose from. Whereas the proximity strain of agglomeration economics highlights the immediate benefits of smart people who are near each other, matching supports the advantages of a broad diversity of opportunities. Matching includes the ability to trade across specialties, whether in employment or in the goods and services offered, and applies to knowledge workers as well as service workers—for example, a clerical worker, janitor, security guard, barista, or rideshare driver. These advantages of urban agglomeration economies require a certain critical mass, and cities and metropolitan regions have a much easier time providing the requisite diversity than do small local governments or rural areas.

Richard Florida (2002, [2012] 2017) famously expanded on this approach to urban growth by arguing that attraction of the "creative class"—a category that includes the well-educated and others with specific skills and interests suited to the modern knowledge-based economy—was essential to urban revitalization and growth. Florida focused on people working in intensely creative occupations such as science, the arts, architecture, and writing, and in knowledge-intensive fields like financial services and high technology. To attract them, Florida argued, cities should offer amenities and a cultural climate—including "tolerance" and diversity—that appeals to young, upwardly mobile, and geographically mobile professionals (Florida [2012] 2017, 244–249). Many of the most populous and fastest growing regions, Florida posited, were distinguished by a new model of economic development that takes shape around what he referred to as the three Ts of development—technology, talent, and tolerance—with the most successful metropolitan areas excelling at all three (Florida [2012] 2017, 228–236).

For Florida, the attraction of the creative class to a city or metropolitan region would have significant positive spillovers for an entire metro region, as the concentration and interaction of creative people spurred high levels of innovation and the expansion of technology-intensive sectors in the region. The growth of the creative class in a metropolitan region would also, he argued, lead to the growth of the "service class," because the service economy is in large measure a response to the demands of the creative economy (Florida [2012] 2017, 146–148). This multiplier effect from the growth of the creative class would make these regions more economically resilient over the long term (Florida [2012] 2017, 50–51; Moretti 2012, 58–63).

Florida has had his share of critics, some of whom questioned the causal relationship between the presence of the creative class and economic growth (Rausch and Negrey 2006). Despite the criticism, many cities embraced urban revitalization plans and economic development policies that mirrored the "creative" city vision that Florida promoted. They aimed to provide cultural amenities and high levels of local service to attract and retain this class of mobile urbanites and adopted innovative financing and development strategies that would attract creative talent and firms. There is evidence to suggest that those kinds of campaigns worked. Mobility patterns suggest that the educated, skilled, and talented mobile residents are disproportionately drawn to a small number of "cool" cities and concentrate in those surrounding regions (Frey 2011).

The most successful urban metropolitan regions—for example, those that surround US cities such as New York, San Francisco, Boston, and Chicago and European cities such as London, Paris, Milan, and Barcelona—have attracted successful industries that have done well in the information economy, have high levels of educated and skilled residents, and provide a rich set of consumption activities such as theater, museums, and restaurants, among other attributes (Glaeser et al. 2001; Frey 2012). In contrast, many older, dense urban regions surrounding cities such as Detroit and St. Louis in the US and Manchester in the UK have industries that have done less well, lower levels of highly educated and skilled workers, and few consumption advantages (Gleaser et al. 2001). Some like Pittsburgh and Detroit have bounced back or are on track to do so, largely through a strategy aimed at providing the type of amenities that attractive the creative class (Frey 2012). The Pittsburgh region, for example, was rated one of the five best places for the "creative class," in part by investing in arts institutions and sports venues and by transforming its old industrial area into an entertainment and shopping destination (Davidson and Foster 2013, 99). Similarly, Detroit's Creative Corridor Center (DC3), opened in 2010 by the local government, has invested in arts and cultural institutions, has fostered the creation of new enterprises and opportunities for designers, and is developing the next generation to follow careers in the creative industries. Despite the development of its Downtown and Midtown areas, the latter around some of its premier university and hospital institutions, Detroit still has the lowest share of creative class workers among US cities where these workers are agglomerating (Florida 2019).

Even if one fully credits the economic logic of urban agglomeration economics as an explanation or driver of urban growth, its costs have become clear. Richard Florida's more recent work on the "new urban crisis" has tracked the geographic segregation of cities to which the creative class has flocked and has found that affluent, highly educated, and skilled populations tend to cluster in and around central business districts and urban centers, transit hubs such as subway, cable car, and rail lines; universities and other knowledge institutions; and natural amenities such as coastlines and waterfront locations (Florida 2017b). He has argued that the most economically successful cities—particular "superstar cities" like New York and London and "tech hubs" like San Francisco—are also the most unequal.

Middle-class neighborhoods are all but fading away. Suburbs, he points out, are growing statistically poorer than urban areas and "large swaths of them are places of economic decline and distress." These class divisions "form a patchwork of concentrated advantage and concentrated disadvantage that cuts across center city and suburb alike" (Florida 2017b, 7).

This new urban crisis requires, in Florida's view, a "new and better urbanism." What this new and better urbanism looks like is up for grabs, as his long list of policy prescriptions suggests. These prescriptions include rebuilding the middle class by investing in infrastructure, building more affordable housing, reforming zoning laws to incentivize density, fostering mass-transit-oriented development, investing in people through a universal basic income, and developing new urban policies at the federal level, among others (185–218). Some of these proposals require national level action, and others are focused on mayors and other local officials who would implement them. It is hard to disagree with many of them.

At the same time, it is difficult to imagine Florida's set of policy prescriptions displacing the idea that the city is a "location market" in which the benefits of agglomeration or proximity to certain classes and types of people are captured by those best able to grab and retain them (Rodriguez and Schleicher 2012). Increased government policies and regulations are said to be disruptive to this market, creating inefficiencies in socially optimal locations of individuals and land uses that result from relatively unconstrained individual choices. It is this very idea of the city as a market that has imposed costs on so many urban communities by fostering real estate speculation, rising rents, gentrification, and unprecedented expulsions (Sassen 2014). To grapple with the urban crisis that has resulted in the market working as it should requires something more than tinkering with policy around the margins of the market. It requires a rethinking of the city and what and whom it is for.

THE CITY AS A PLATFORM: TECHNOLOGY-ENABLED SMART CITIES

The vision of a *smart city* presents a unique opportunity to innovatively tackle significant urban problems while reinventing the city in a more open and innovative form through distributed data and technological capacity.

The idea of a smart city emerged as a strategy to mitigate problems generated by urban growth and uncontrolled urbanization processes, promising to transform urban life, urban planning, and city hall (Townsend 2014). It is poised to be responsive to a host of civil, social, economic, and ecological problems in cities by deploying information and communication technologies (ICT) and the use of big data to address them (Goldsmith and Crawford 2014).

Smart city technology can be deployed to catalyze economic development, monitor pollution and energy use, adjust traffic patterns to avoid congestion, enable better predictive policing, and deliver better health care and education, among other aspects of city life. In a smart city, high-speed wireless and broadband connectivity is provided by the city as a public good, reaching all communities and populations. Smart devices, sensors, and other technological tools are disseminated throughout the city to enable real time data. processing, management, and analysis.

For some, smart cities have the potential to turn the city into a kind of civic laboratory, enabling and facilitating data-led strategies by integrating design and community-based solutions (Townsend et al. 2010). City leaders can use these tools to enable city government to be more responsive to and engaged with its citizens, including the most marginalized and powerless. In this vision of the smart city, technology can be adapted in novel ways to meet local needs by putting urban residents in "the driver's seat" where they will be able to "respond to subtle social and behavioral clues from their neighbors about which way to move forward" and to "use their distributed intelligence to fashion new community activities, as well as a new kind of citizen activism" (Ratti and Townsend 2011, 42–43).

Smart cities are as varied as cities themselves, in part because of the capaciousness of what the term *smart city* represents. As Robert Hollands (2008) has argued, the term obscures as much as it reveals. A smart city refers to "quite a diverse range of things," including information technology, business innovation, governance, communities, and sustainability (Hollands 2008, 306). Hollands identified two of the main aspects of designated smart cities: the use of new technologies throughout their cities and a strong pro-business/entrepreneurial state ethos. Nevertheless, he noted that many smart city agendas were also concerned with high-tech and creative industries—such as digital media, the arts, and the cultural industries more generally—as well as "soft infrastructure" that includes, for example, knowledge networks, voluntary organizations, safe crime-free environments, and a lively after-dark entertainment economy (Hollands 2008, 309). Still other smart city developments, he noted, manifest a concern with both social and environmental sustainability. Whereas social sustainability fosters social cohesion and inclusion, environmental sustainability focuses on the ecological and "green" implications of urban growth and development (Hollands 2008, 310).

Over the last decade or so, the smart city vision and agenda has expanded and grown, as have the number of global smart cities influencing the best practices that other cities might replicate. One recent study identified twenty-seven leading smart cities around the world—mostly "capital" or "alpha-world" cities in Asia, Europe, and North America—that are focused on "multiple dimensions" of the city beyond just infrastructure and technology (Joss et al. 2019). The authors identified the "discourse" of these smart cities—that is, how they describe their smart city—along several dimensions. Reform of urban governance, including systems integration and broader collaboration across society, was a centralizing theme of smart cities, followed to some degree by a focus on international global activity. The picture that emerges, according to the study authors, is that the smart city "is seen as an opportunity to embark on fundamental infrastructure modernization activities, for which appropriate governance mechanisms are called for" (Joss et al. 2019, 13).

Today, smart cities are the products of a consortium of aligned actors, which includes an epistemic community (a knowledge and policy community), an advocacy coalition (of stakeholders and vested corporate interests), and embedded technocrats in local government (e.g., chief technology and information officers, chief data officers, data scientists, smart city specialists, and IT managers) (Kitchin et al. 2017). At the same time, as Dan Kitchin et al. (2018) explain, the "focus, intention, and ethos of smart city ideas, approaches and products remains quite fragmented and often quite polarized." On one hand are smart city enthusiasts—scientists, technologists, technocrats, companies, and government—who want only to develop and implement the technologies and initiatives to improve cities and city life "often with little or no critical reflection on how they fit into and reproduce a particular form of political economy and their wider consequences beyond their desired effects." On the other hand, are critics who raise a host of concerns rooted in political, ethical, and ideological perspectives—focused on issues of power, equality, participation, labor, surveillance, and other concerns—who come largely from the social sciences (geography, urban studies, sociology, etc.) and civic organizations. Although their critiques are powerful, they often "provide little constructive and pragmatic (technical, practical, policy, legal) feedback that would address their concerns and provide an alternative vision of what a smart city might be" (Kitchin et al. 2018).

The tensions embedded in the ideation of the smart city can manifest in a lack of trust between public authorities and the communities who are seen as its beneficiaries. The recent failure of a technologically sophisticated, state-of-the-art, sustainable neighborhood project on Toronto's waterfront created by Google subsidiary Sidewalk Labs is illustrative of the contradictions and pitfalls of the best-laid smart city plans. Unable to trust the intentions of the sensor-based surveillance and data-driven "responsive" service and frustrated by a lack of transparency about the scope of privacy and data protection, Toronto residents lost faith in the project, and it collapsed. The heavily celebrated project, with a sense of inevitability, ultimately represented a failure of urban governance. As Ellen Goodman and Julia Powles (2019) argue in their in-depth analysis of the project's demise, its failure is not attributable to the public's grievance with technology or innovation per se. Rather, its failure is attributable to the fact that the public authority, a partner on the project, lost the confidence of the public that the project's vision was compatible with democratic processes, sustained public governance, or the public interest (Goodman and Powles 2019). The public evinced deep skepticism with the "centrality and hugely asymmetric power of a private corporate group" exerting dominance over nearly every aspect of the future district (Goodman and Powles 2019, 498).

It is possible, as Duncan McLaren and Julian Agyeman (2015) have proposed, to reorient smart cities by "harnessing smart technology to an agenda of sharing and solidarity, rather than one of competition, enclosure, and division" (McLaren and Agyeman 2015, 5). For McLaren and Agyeman, "sharing and cooperation are universal values and behaviors," and if cities are shared creations with shared public services, streets, mass transit, and shared spaces, "truly smart cities must also be sharing cities" (24). Cities that embrace the smart/sharing city paradigm would be inclined to expand and share physical and data infrastructure more widely and put idle public resources to use in creating an inclusive urban economy (McLaren and Agyeman 2015, 71–77).

The idea of the shared, smart city offered by McLaren and Agyeman is consistent with our idea animating the co-city—that the city itself should be conceptualized as a shared resource or shared infrastructure. Like us, the authors embrace collective forms of resources sharing, peer-to-peer production, and co-produced goods and services. They also have as a goal the sharing of the "whole city," including its technological and digital infrastructure, toward particular normative ends (McLaren and Agyeman 2015, 5) This is contrasted with a technocratic and market-driven vision of a smart city that ignores questions of power and distribution in the accessibility of basic goods and services in contemporary urban environments. The sharing paradigm, as they construct it is situated in contemporary theories of "just sustainability" and the human capabilities approach that are drawn from a number of classic and contemporary philosophers such as John Rawls, Michael Sandel, Amartya Sen, and Martha Nussbaum (McLaren and Agyeman 2015, 199–208).

What the sharing city paradigm lacks, however, is a more refined understanding of what kinds of sharing practices and policies would satisfy its normative aims and which kinds would fall short. By the authors' own account, "sharing" practices in the cities that they highlight fall along a spectrum that ranges from the commercial to the communal, from city-wide to informal neighborhood practices. Cities like Amsterdam and Seoul, for example, have embraced both the "smart" and the "sharing" city labels, using technology to empower residents through open-access public data and free platforms for citizen participation, and to catalyze citizen development of sharing economy apps, enterprises, and start-ups. Other sharing cities have embraced policies or exhibit practices that leverage their city assets and public buildings to support community-based sharing enterprises and organizations, toward social and economic inclusion. Still others embrace policies that facilitate more profit-driven sharing economy platforms like Airbnb and Uber that are in tension with communal or solidarity forms of sharing.

For cities whose practices are more aligned with the authors' vision of a sharing city, many questions remain unanswered. For instance, how might these practices scale and replicate in different urban contexts? What are the informal and formal mechanisms that residents and communities utilize to share or *co-produce* common goods with other actors, and what are the challenges they face given the political economy of cities and the market forces that constrain these actors? What is the role of the state in facilitating sharing practices across a city that are embedded in communities' material needs and differential capacities? What kinds of place-based governance (or institutional) arrangements constitute best practices for sharing? What does it mean to consider the whole city a shared resource, and to co-create or co-produce goods and services that are accessible and open to those most in need of them?

Answering these questions involves a more comprehensive assessment of city policies and practices than that offered by McLaren and Agyeman. For this reason, the co-city framework that we have developed is rooted in significant part in a multiyear empirical study surveying hundreds of policies, practices, and projects in different cities around the world to enhance our understanding of the various ways that built, environmental, cultural, and digital goods are being co-created and co-governed in different geographic, social, and economic contexts. The empirical project sought to obtain, from on-the-ground examples, recurrent design principles and common methodological tools employed across the globe and for different urban resources. We have extracted the characteristics of these diverse efforts to develop a framework that reflects the conditions and factors that we observe as necessary to rethink the city as a shared infrastructure on which a variety of urban actors can cooperate and collaborate and in which various initiatives of collective action can emerge, flourish, and become sustainable.

What kinds of resources should be shared, collaboratively governed, or held, and which actors can (or should) manage them are in part applied questions that can be answered only by reference to the specific location and context of each city. For this reason, we conceptualize and frame the co-city as a form of urban experimentalism guided by a set of design principles that can be adapted to local context.

THE CITY AS A COLLECTIVE GOOD: RECLAIMING THE RIGHT TO THE CITY

The idea of the "right to the city" was introduced in the scholarly debate by the philosopher Henry Lefebvre ([1968] 1996) in his examination of the urban roots of social movements. As articulated by Lefebvre, the right to the city is a framework through which citizens can reclaim or re-appropriate city space, inhabit and share its spaces, and actively participate in formation and stewardship of city space. As Mark Purcell has argued in his close reading of Lefebvre, the right to the city should be interpreted, at least in part, as a struggle to "de-alienate" urban space and to "reintegrate" it into the web of social connections among urban inhabitants, activating inhabitants to participate in the collective stewardship of urban life and to manage the production of urban space themselves (Purcell 2013, 150). The right to the city is rooted also in the struggle between exchange value and use value, or between the city as a site of accumulation and the city as an inhabited place that nurtures the use value and needs of its inhabitants (Purcell 2013, 150).

Critical urban geographer David Harvey considers the right to the city as a fundamental but neglected *human right* to "make and remake the world that we live in" and the "right to change and reinvent the city" by those whose labor produces and reproduces the city (Harvey 2012, 4, 137). For Harvey, the right to the city idea embodies far more than a right of individual or group access to the city's resources. Reinventing the city also requires endowing urban inhabitants with the "collective power" over the processes of urbanization and in decisions about urban space (Harvey 2012, 137). Harvey builds on Lefebvre's vision of "urban" as a process rather than a fixed space or set of resources.

Despite the elegance of the theory and various forms of articulation, there are uncertainties and contradictions about what exactly the *right* to the city entails in practice. For one, we might locate the right as access to the city's physical infrastructure, as some progressive property scholars suggest. Nicholas Blomley, for example, powerfully argues for the right of the poor "*not* to be excluded" from the property of the city (Blomley 2008, 320). He observes that "we can find many examples in cities across the world where state or private actors use the power to exclude, which is central to private property, to displace, evict and remove the poor" (Blomley 2008, 316). This is a call for recognition on behalf of the poor of a collective claim to neighborhoods and structures within them as a response to the appropriation and enclosure of those places in ways that exclude the poor from cities. This collective claim is a highly *localized* one that includes streets, parks, and buildings, among other resources, over which the poor have legitimate interest as both a symbolic and a practical matter (Blomley 2008, 316).

The right to the city could include the right to collective political power as it relates to public deliberation and participation. The right to collective political power entails, at the least, that urban inhabitants should have an increased voice in local decision-making processes and exercise greater control over the forces shaping city space. In other words, the right to the city must mean, as its adherents agree, the right to *governance* of the city by its inhabitants. Lefebvre is clear that the decision-making role of *citidans*—urban inhabitants—must be *central*, even if he is not explicit about what exactly that centrality would mean in practical terms, including whether decisions that produce urban space should be made entirely by urban inhabitants (Purcell 2002). The right might include, at the least, the right to reject unjust collective decisions taken by local authorities (Attoh 2011).

Notwithstanding its lack of granular specificity, the right to the city discourse and framework have found practical application in some Latin American and European contexts. Most notably, in 2001, Brazil incorporated the right to the city into its City Statute, a federal law regulating urban development under Brazil's 1988 Constitution. The City Statute sets out general guidelines that must be followed by federal, state, and local governments to ensure "democratic city management," including through mandating participation in planning processes and adopting the principle of "the social function of property and the city." The "social function of property" principle is found in many constitutions around the world, particularly in many Latin American countries. The doctrine embraces most broadly the idea that an owner cannot always do what she wants with her property; rather she is obligated to make it productive, which may include putting it at the service of the community (Foster and Bonilla 2011). In other words, sometimes the state is obligated to require individuals to sacrifice some property rights in order to put property to its productive and socially functional use, or to do so itself. Through several new legal instruments allowing municipalities to control and expropriate land, the City Statute established that the development of urban land (in either the formal or the informal sector) and buildings should be determined first and foremost by its social "use value" over its commercial "exchange value" (Fernandes 2007).

Similarly, Mexico City passed its Right to the City Charter in 2010, setting out six fundamental principles that incorporate human rights and a "collective right" of urban inhabitants to the city. These principles and rights include the social function of the city and of property, participative management and democratic production of the city, sustainable and responsible management of its commons and resources, full exercise of human rights in the city, and equitable right to enjoy the city itself. The drafting of the charter involved a bottom-up process by the Urban Popular Movement (Moviemento Urbano Popular, MUP) with the participation of over 3,500 citizens through consultations and public meetings.

Both the Brazil City Statute and the Mexico City Charter were groundbreaking in instantiating the right to the city as a legal and governance principle, although both have faced challenges in implementing their broad rights and guarantees (Fernandes 2011; Friendly 2013). Market pressures have made it difficult to sustain some of the social housing and other welfare provision gains on behalf of the urban poor, given the rise of urban land values in Mexico City, for example, leading to displacement and expulsions of the urban poor (Adler 2017). Discontent with the implementation of the promise of right to the city through legislation and policy has led the precariat (the lowest-income class) to occupy vacant and underutilized land and buildings for housing (Irazábal 2018). It has also led to a push toward more collective or cooperative forms of ownership in the place of government-subsidized individual property that has faltered in the face of speculative urban property markets (Adler 2017).

More recent citizen-organized *rebel city* platforms have emerged in the aftermath of frustrated efforts to implement legal reforms like those in Brazil and Mexico City. David Harvey popularized the concept of rebel cities to encourage urban inhabitants to take an active role in resisting the process of capital-intensive urbanization—a "perpetual production of an urban commons (or its shadow-form of public spaces and public goods) and its perpetual appropriation and destruction by private interests" (Harvey 2012, 80). In a rebel city, urban inhabitants actively engage in the struggle for reclaiming their right to the city that is under the attack by predatory capitalist forces, to retain the value that they collectively produced. Harvey highlighted urban revolutionary movements such as urban protests and sit-ins in London, Madrid, and Barcelona and the Occupy Wall Street movement in New York, as examples of the potential of rebel cities.

The turn toward rebel cities has taken hold in cities like Barcelona and Naples to strike more directly at the outsized role that capital and market forces play in controlling urban land and other critical resources. Barcelona's transformative citizens' electoral platform, *Barcelona en Comú* (Barcelona in Common), has successfully pushed for progressive local policies on housing and energy provision and advanced the right to information and to open, participatory decision making through new digital and platform technologies (Charnock et al. 2019). In the city of Rome, organized residents proposed a Charter of Common Rome identifying ten fundamental principles aimed at the recognition of the right to use city infrastructure and vacant spaces. This charter defines these ten principles for the participatory management of the public goods of the city, including the inalienability of state-owned assets, the right to "common use" of such assets, the recognition of the urban commons, and the right of citizens to co-manage the urban commons and participate in decision-making processes related to them (Decide Roma).

Our co-city framework shares much in common with the right to the city and rebel city approaches. Like the right to the city vision, our framework is rooted in a collective claim to certain public spaces, vacant land, and abandoned structures as shared, common resources. Conceiving the co-city through the lens of the right to the city also requires conceptualizing urban governance along the same lines as the right to the city—the right to be part of the creation of the city by participating in the stewardship or governing of urban resources. From a normative perspective, the co-city framework brings squarely into view questions of social and distributive justice. As a matter of distribution, the resources of the city should be shared more widely throughout its communities and on behalf of its inhabitants, particularly the most vulnerable and those subject to what Saskia Sassen (2014) calls "expulsions"—unprecedented displacement, evictions, and eradication of living spaces and professional livelihoods. We join progressive property scholars, like Nicholas Blomley (2008), who poignantly call for the recognition on behalf of the poor of a collective claim to neighborhoods and communities as a necessary response to the appropriation and enclosure of urban spaces and infrastructure through private property rights.

It is communities, as property scholar Greg Alexander (2009) has argued, that are the mediating vehicles through which people acquire the resources that they need to foster the capabilities necessary to function and flourish. In other words, human flourishing requires resources, although ownership of those resources is not always required for all kinds of capabilities; use and access may be enough (Alexander 2020). Nevertheless, as progressive property scholars argue, "however the details are conceived, attention to human beings' social needs pushes strongly in the direction of a state obligation to take steps to provide substantial and realistic opportunities for people to obtain the property required for them to be able to participate at some minimally acceptable level in the social life of the community" (Alexander and Peñalver 2009, 148).

Consistent with the capabilities approach to questions of justice, in the co-city framework the state—the central or higher-level government—plays a crucial role in enabling and facilitating collective action as well as providing some of the necessary resources to generate and sustain resources as urban commons. The co-city approach federates a wide spectrum of actors, agents, and sectors in the city, including single city inhabitants or informal groups, civil society organizations, knowledge institutions, and private institutions, to pool resources in order to co-govern or steward city infrastructure, assets, networks, and services. The result of these *pooling* practices is the co-production and co-governance of affordable housing, land for growing food, green or recreational space, shared entrepreneur and workspaces, and new forms of broadband connectivity and energy provision. These pooling practices—what we also refer to as *pooling economies*—emerge out of the collective action of different actors who, using existing urban infrastructure, mix and match their resources to expand their capacity to construct and co-govern urban essential resources. The participation and active urban citizenship manifested in these pooling practices resonate with what historian Peter Linebaugh has called "commoning"—social practices of users in the course of managing shared resources and reclaiming the commons (Linebaugh 2008). It also builds on Lefebvre's vision of *urban* as a process rather than a fixed space or set of resources.

THE CITY AS A COMMONS: THE CO-CITY VISION AND FRAMEWORK

The co-city framework is supported by the conceptual pillars of the urban commons and the idea of the city as a commons. Starting over ten years ago, we began to explore the idea that urban infrastructure and other resources within cities could be collectively or cooperatively governed by city residents, most often sharing this governance responsibility with other actors depending on the scale of the resource. Our study of the *urban commons* began as separate projects that each investigated how various kinds of urban assets such as community gardens, parks, neighborhoods (Foster 2006, 2011) and urban services and infrastructure such as urban roads (Iaione 2008, 2010) could be reconceived as common resources. We later joined our efforts to conceive the city itself as a commons, which we defined as a shared infrastructure on which a variety of urban actors can cooperate and pool resources and where various initiatives of collective action can emerge, flourish, and become sustainable (Foster and Iaione 2016). Thinking of the city as a commons is a way to acknowledge that the city is generative, capable of providing for different social and economic needs of its population.

The *commons* has a long historical and intellectual lineage ranging from the enclosure movement in England to the Nobel Prize winning work of Elinor Ostrom. In her groundbreaking work, Ostrom documented the success of human communities around the world that rely on natural resources—such as lakes, forests, and fisheries—to collectively govern those resources by creating "institutions resembling neither the state nor the market," which have had reasonable degrees of success over long periods of time (Ostrom 1990, 1). Under certain conditions, Ostrom found, resource users collectively decide how to produce value from the resource, enforce rules and norms of use, and avoid overconsuming or depleting the resource over time.

Ostrom's work sparked the study of a variety of user-governed, shared resources beyond natural resources that require thinking about the process of developing and enforcing rules, social norms, and other legal or governance tools for sharing and sustainability utilizing those resources. Scholars have conceptualized and articulated new kinds of commons that involve "communities working together in self-governing ways to protect resources from enclosure or to build newly open-shared resources" (Hess 2008, 40). These include knowledge commons, cultural commons, infrastructure commons, and digital commons, among others. Until the last decade or so, there had not been a serious effort to apply the commons to the built environment in cities. Although the literature on natural resource *commons* and *common pool resources* is voluminous, virtually no scholars had endeavored to transpose Ostrom's insights into the urban context in a way that captures the complexity of the *urban*—the way that the density of an urban area, the proximity of its inhabitants, and the diversity of users interact with a host of tangible and intangible resources in cities and metropolitan areas.

Through our individual and joint work, it became clear that cities and the many kinds of urban resources within them differ in important ways from traditional, natural resources commons as well as other kinds of new commons. We needed to start with Ostrom's work and her design principles but also to acknowledge the limits of her framework and its applicability to the urban environment (Foster and Iaione 2019). The economic, regulatory, and political complexity of cities for us means that although it would be tempting to simply transpose Ostrom's work and findings to the city and to apply them to the stewardship and governance of many kinds of public and shared resources in the city, doing so would obscure rather than illuminate any concept of the urban commons. Moreover, our work raised larger social and economic issues related to broadscale urbanization than existing commons literature adequately accounted for. We realized that we needed a different approach that bridged urban studies and commons studies, which encompass multiple disciplines ranging from law and economics to political science and geography (Iaione 2015).

Apart from our work, the urban commons has become an important conceptual framework across many disciplines for examining questions of resource access, sharing, governance, and distribution of a range of both tangible and intangible resources in cities (Borch and Kornberger 2015). Urban commons in this growing body of literature encompasses both material and immaterial resources—ranging from housing, urban infrastructure, and public spaces to culture, labor, and public services (Dellenbaugh et al. 2015). The language of the commons is deployed to disrupt the boundaries separating public and private goods and services in cities and to open up those goods and services to public use in ways that do not depend on and are not controlled by a prevailing authority (Stavrides 2016).

Progressive scholars and activists also invoke the idea of the urban commons to bring under scrutiny the ways that capitalist power has resulted in the enclosure of urban space by economic elites (Harvey 2012). The literature on urban commons in part investigates the city as a site of capital production and surplus and a place of contestation for resources (Stavrides 2016). For these scholars, the urban commons must be "wrenched" from the capitalist landscape of cities out of fear that collective or common resources are always susceptible to being co-opted by the market (Huron 2015). The roots of progressive reformers' commons analysis are traceable to the work of Michael Hardt and Anthony Negri (2009), who refer to the "common" (rejecting the term "commons" as a reference to "pre-capitalist shared spaces that were destroyed by the advent of private property") as the product of shared efforts by city inhabitants. Cities are, as they argue, "to the multitude what the factory was to the industrial working class"; in other words, it is the "factory for the production of the common," a means of producing common wealth (Hardt and Negri 2009, 250).

We embrace the potentially disruptive role of commons discourse to highlight the privatization and enclosure of city space and to interrogate who has access to our shared resources in cities and how they are allocated and distributed. The language of the commons is a powerful counterclaim to resources on behalf of city inhabitants subject to the dispossession and displacement that has resulted from unfettered capital accumulation. Making claims on urban resources and city space as a *commons* creates an opening, or space, to bring under scrutiny the character of particular urban resources in relationship to other social goods, to other urban inhabitants, and to the state. Thinking of some urban assets as resources to be collectively or collaboratively stewarded by an identified community or group of people requires us to move beyond the public/private and market/state binary choices to which we often default in thinking about resource use and control. It is in the space between public and private and between market and state that we locate a set of rich conceptual and practical possibilities.

Locating the rich set of practical possibilities for collectively stewarding and governing urban infrastructure has motivated our development of the co-city framework. In addition to our scholarship, the co-city framework is rooted in our own experiences working with city officials and subsequent empirical investigation of different policies and practices in cities around the world. We began applying our conceptual approach to the urban commons and to the city as commons in Bologna, Reggio Emilia, and Rome Italy, as part of their experiments to create a *collaborative city*.

In 2011, the local administration in Bologna began a process to put into place a set of policies that would reshape the social, economic, and political functioning of the city. Two of the centerpieces of this effort, in which we participated, were the drafting of two new local regulations. The first regulation concerned "The Realization of Micro-Projects of Improvement of the Public Space by the Civil society." The second regulation, which became more well known, was on "Collaboration between Citizens and the City for the Care and Regeneration of the Urban Commons."

The second Bologna regulation, adopted in 2014, empowered the local administration to enter into "pacts of collaboration" with some mix of city inhabitants, representative civic groups, local nonprofits, and local businesses. The pacts are created through a co-design process that includes robust public participation, resulting in an agreement that describes the urban resource that is the subject of collaborative regeneration and/or co-management and the project scope including the duration and the respective roles and commitments of the actors involved. The regulation also provides for different forms of fiscal, logistical, training, and organizational support from the city to realize the goals and implementation of the pacts.

The regulation on collaboration between citizens and the city was the cornerstone of this urban economic transformation and civic engagement process, but it was only a part of the larger experimentation. The regulation was designed to rely on neighborhood experimental projects as its starting point. These were fieldwork activities that consisted of three *governance experimentation labs*, which comprised a mentoring and co-design program in which local officials worked together with local NGOs and neighborhood residents with the support of experts and scholars. In these governance labs, communities of actors were able to identify and co-design projects that would revitalize or regenerate three types of *urban commons*—cultural assets, green spaces, and city-owned buildings—and to overcome legal and procedural obstacles that could hinder meaningful cooperation with local officials and more cooperative engagement of neighborhood residents.

Following the successful adoption and implementation of the Bologna regulation (to date, around six hundred pacts of collaboration signed and implemented), other cities began experimenting with similar regulation and policies. First, several Italian municipalities embraced the Bologna approach, sometimes copying verbatim the Bologna regulation and other times instituting their own innovative approaches to fostering co-governance of urban resources. On the heels of the Bologna regulation, for example, the city of Reggio Emilia enacted the "Neighborhood as Commons" policy, implemented through "citizenship agreements," which spurred hundreds of innovation projects in neighborhoods that served almost 14,000 users and were co-designed in citizenship labs. The city of Naples, as another example, adopted a "civic and collective urban uses" policy that recognizes informal management by city residents of city-owned buildings. The city of Turin approved a regulation that blends the Bologna and Naples approaches through "civic deal(s)" that recognize and grant rights of collective use, management, stewardship, and ownership of shared urban assets specifically designed to reduce poverty in the city. In Rome the regulatory and governance complexity suggested the creation of a citizen science platform and project called Co-Roma, which allowed the experimentation in a large metropolis of adaptive and polycentric governance mechanisms such as creation of a community cooperative for vulnerable neighborhoods, a city-region-citizens contract for the Tiber River, and a house of emerging technologies for sustainable development.

Subsequently, other cities in Europe adopted analogous initiatives and regulatory approaches. Madrid passed an ordinance on social cooperation for the urban commons, and Barcelona is implementing a "Citizen Asset program for community use and management." More recently, cities in northern and eastern Europe—Ghent (Belgium), Amsterdam (Netherlands), Gdansk (Poland), Presov (Slovakia), and Iasi (Romania)—are working on a similar, common regulatory scheme under the auspices of the *Civic eState Urbact project*. We discuss the Bologna project and some of these other examples from European cities in chapter 3.

Similar projects have since emerged in North American cities, including cities as diverse as New York City and Baton Rouge (Louisiana), and in the Global South cities such as San Jose (Costa Rica) and Sao Paolo (Brazil). These projects are less focused on regulations and ordinances, and more on the co-creation of innovative forms of affordable housing, broadband and wireless networks, and regeneration of public spaces or vacant land and buildings, particularly in marginal or disadvantaged communities. For example, in New York City, the co-city approach is using the infrastructure of a *smart city* to create a community governed broadband network in Harlem to bridge the digital divide there that leaves one-third of households and families without access to broadband internet at home. In Baton Rouge, the co-city approach is deployed to revitalize a historically African American four-mile commercial corridor developing a portfolio of innovative community-based institutions for resident stewardship and governance of existing community assets. We discuss these two projects in chapters 4 and the conclusion, respectively.

On the basis of the Bologna experience and the interest of other cities and communities in adopting similar regulatory or public policy approaches to co-management and co-governance of urban infrastructure, we decided to launch an empirical investigation of the ways that collectively shared and collaboratively stewarded resources can be created and sustained in different political, social, and economic environments. To date, we have mapped over two hundred cities around the world and over five hundred policies and projects within them as part of the co-cities project. The data set, contained in an open book, published on the web platform (commoning.city), and summarized in the appendix to this book, provides case studies of projects and public policies from the cities mapped. From those cities mapped, we more closely analyzed, through interviews with relevant stakeholders and/or more extensive desk research, 140 cities with 289 cases within them.

The cities that we surveyed and analyzed were selected on the basis of the existence of a project or policy relevant to creating, enabling, facilitating, or sustaining collaboratively or cooperatively shared resources utilizing the existing infrastructure of cities. To capture diversity, we identified and included a group of case studies for every geographical area, including southern Europe, central and northern Europe, eastern Europe, North America, Central America, Latin America, northern Africa, sub-Saharan Africa, Asia, and Oceania.

The examples discussed throughout the book are based in large part on the most robust case studies developed from our empirical exercise. These roughly thirty exemplary cases offer important examples and insights from cities worldwide in which there are emerging community or city-level initiatives that enable and facilitate the pooling of resources that result in urban goods and services that are collectively governed, stewarded, and shared by marginal and disadvantaged populations. These increasingly taking the form of city policies supporting community control of neighborhood land and physical infrastructure, new forms of co-housing, limited equity cooperatives, and community-shared digital networks, among others.

The goal of the empirical aspect of the co-cities research project has been to extract some of the characteristics of these diverse efforts to develop a common framework and understanding of recurrent principles and common methodological tools employed in different contexts and for different urban resources. The result of this research project is to offer those observations as *design principles* to help guide the experimentation of the co-city approach beyond the examples we offer in the book; the design principles can be adapted to local context.

We have distilled five basic design principles, or dimensions, from our practice in the field and the cases that we identified as sharing similar approaches, values, and methodologies. These five key design principles of co-cities are the following:

- Principle 1: *Collective governance (or co-governance)* refers to the presence of a multistakeholder governance scheme whereby a local community emerges as an actor and partners (through sharing, collaboration, cooperation, and coordination) with four other possible categories of urban actors to co-produce and/or co-govern urban resources; the four actors include public authorities, private commercial entities, civil society organizations, and knowledge institutions such as schools, universities, libraries, cultural institutions, museums, and academies.
- Principle 2: *Enabling state* expresses the role of the state (usually local public authorities) in facilitating the creation of shared urban resources

and supporting collective governance arrangements for the management and sustainability of these resources.
- Principle 3: *Social and economic pooling* refers to the presence of autonomous, self-sustaining institutions (e.g., civic, financial, social, or economic) that are transparent, collaborative, and accountable to local communities and operate within nonmainstream economic systems (e.g., cooperative, social, solidarity, circular, cultural, or collaborative economies) that pool resources and stakeholders toward the creation of new opportunities (e.g., jobs, skills, and education) and services (e.g., housing, care, and utilities) in underserved areas and neighborhoods of the city or for vulnerable inhabitants.
- Principle 4: *Experimentalism* is the presence of an adaptive, place-based, and iterative approach to urban planning, legal reforms and policy innovations that enable the co-creation of collectively shared urban resources.
- Principle 5: *Tech justice* highlights access, participation, and co-management and/or co-ownership of technological and digital urban infrastructure and data as an enabling driver of cooperation and co-creation of shared urban resources.

We describe these principles in more detail in chapter 5. The appendix illustrates that even in our exemplary case studies, the presence of each of these principles varies. How strongly each is present in a particular case can depend on different contexts and the kinds of resources being constructed and shared. The design principles are not intended to be prescriptive but rather a starting place to create conditions that reflect those principles. Throughout our examples, drawing on our own experiences, we have also been attentive to the recurring legal, financial, institutional, and digital/technological tools and mechanisms that contribute to the presence of the design principles. Creating the conditions for a co-city requires attention to its aims, attention to its basic principles, and a willingness to learn from the experiences and experiments of others who are pioneering this approach.

THE CHAPTERS

This book has been designed to offer the reader a theoretical and conceptual map to the co-city framework as well as real-world examples of how shared goods are constructed from available and accessible urban infrastructure. The book then introduces and analyzes the emergence of pioneering legal and policy responses that facilitate collective governance or stewardship of resources in different kinds of cities around the world.

Chapter 1 introduces the foundational challenge from which our framework emerges: the tension between the exchange and use value of urban land and infrastructure. Both in resurgent cities like New York and in *minimal* cities like Detroit, residents in communities that lack basic goods like affordable housing or internet access and lack adequate employment opportunities want access to urban infrastructure—vacant and underutilized land and structures—to transform them into affordable housing units, urban farms, or spaces for local entrepreneurs or artists. Local governments most often value these assets for their potential market exchange value. Some even view the divestment and sale of vacant public property as an economic necessity. At the same time, these assets are valued by residents for their relationship to the community, whether on the scale of a block or of a neighborhood. The use value of these resources comes from their everyday use, the solidarity that it creates among its users, and its accessibility to surrounding residents. Communities living near these assets may endeavor to work with the public sector and other actors to construct new urban resources or goods to meet the needs of surrounding communities: housing, parks, urban farms, co-working spaces, and other resources and goods.

Our starting point is to rethink the city, and specifically the infrastructure of the city, as a shared or common resource that is capable of being generative through the collective action of various urban actors who can construct new goods from this infrastructure to meet the social and economic needs of urban populations. In this sense, we can think of the city as a *commons* and recognize as legitimate, and even innovative, the efforts by residents to utilize land and other infrastructure and to pool resources with other actors in order to construct informal neighborhoods and settlements, community gardens and urban farms, mesh wireless networks, and new limited equity housing and commercial spaces that are collaboratively governed by community, public, and private participants for long-term affordability and sustainability. The chapter reflects on the role of the state—central authorities—in enabling and facilitating these efforts throughout a city, creating a polycentric system of urban governance.

Chapter 2 delves deeper into the concept of the urban commons, beginning with understanding how urban commons differ from the kinds of common pool resources that Elinor Ostrom and others have studied. Some urban resources, such as parks and urban gardens, at first glance may resemble the kind of natural resource commons that are the subject of Ostrom's (and many others') work. However, we argue that many facets of collectively stewarded or governed urban resources are notably distinct when observed and studied in the context of contemporary cities that are often crowded, congested, socially diverse, economically complex, and heavily regulated. As such, we highlight some characteristics of *constructed* urban commons that are not captured well in the literature on user-governed natural resource commons. We identify three elements that are key to the creation of many urban commons and that are not always present in collectively managed natural resource commons. These are the role of central authorities (the state) in enabling the creation and sustainability of urban common; legal and property experimentalism or adaptation; and social and economic *pooling*.

We also draw a clear distinction between top-down and bottom-up urban commons. The former kind are exemplified by park conservancies and business improvement districts. Although these institutional arrangements resemble some of the features of Ostrom's design principles for collective governance of shared resources, they are not the kinds of constructed commons that our work has identified as engaging resource users in the stewardship of shared common goods. Instead, they represent the kinds of self-professed *public-private partnerships* that can carry costs for urban communities least able to participate in the stewardship of the common resources that they manage. Many other kinds of urban commons emerge, on the other hand, from bottom-up efforts of residents or resource users who are motivated to overcome traditional collective action problems and to collaborate to construct new goods and services that many urban communities lack or find inaccessible to them. These constructed commons are increasingly taking the form of community land trusts, new forms of co-housing, limited equity cooperatives. We discuss the way that these institutions can become nested within the institutional framework of the city through public-community partnerships or public-private-community partnerships and can scale with the support of local policies and public resources to create a polycentric network of urban commons in the city.

In chapter 3, we turn to the emergence of city policies that enable, facilitate, and support urban commons and allow them to nest within the governance infrastructure of the city. We examine the emergence of public policies in a handful of cities that endeavor to deeply engage citizens through public-public and public-community partnerships with the goal of implementing an arrangement in which citizens are governing the city rather than merely being governed. The policies described in this chapter situate the local government as an enabler and facilitator of collaboration and ultimately of political and economic redistribution through shared urban goods and infrastructure. While communities and other stakeholders organize themselves autonomously as potential collaborators that can collectively manage urban resources, city officials and staff are tasked to assist, collaborate, and provide technical guidance, which can include data, legal advice, communication strategy, design strategies, sustainability models, and other assistance, to those efforts. The governance output that emerges from implementation of these policies is the co-design of a variety of urban commons as well as the co-production of community goods and services at the city and neighborhood level.

To better understand and turn a critical eye toward these policies, we organized them in two categories: *declaratory* versus *constitutive* policies or laws. A declaratory policy acknowledges the existence of collectively managed individual resources or neighborhood institutions as forms of urban commons. These policies officially recognize the right of these communities to self-organize and might entail recognition of social norms agreed upon by the community and/or validation of the public value produced by the community that justifies their right to utilize the shared resource. The local government might even enter into some sort of agreement with the collective, lending legitimacy and some stability to the effort as well as indirectly encouraging other bottom-up efforts throughout the city. Constitutive policies, on the other hand, embody a more top-down, institutionalized approach. They are specifically aimed at encouraging the creation of urban commons throughout the city and endeavor to create the conditions for governing some city resources collaboratively by offering new legal authority or adapting existing laws. Those two approaches are implemented through a range of different legal tools ranging from collaboration pacts or agreements to civic-use regulations allowing the private use of public assets. Both approaches present ongoing challenges and attendant costs, which we discuss in the chapter.

Chapter 4 conceptualizes the urban co-governance that is reflected in policies and settings in which communities interact with the state and other actors to collectively create and steward urban resources like land, buildings, and even utilities and wireless networks. This urban co-governance embraces the role of the facilitator state, in which city officials and staff are tasked to assist by providing resources and technical guidance to help create the conditions for co-governance, sometimes in the form of public-public and public-community partnerships. It also creates a system that at its core redistributes decision-making power and influence away from the center and toward a network of engaged urban actors. The co-governance model that we embrace takes as a starting point the active involvement or participation of urban residents in the management and governance of urban resources to support the livelihood and well-being of their communities. We argue, however, that to truly generate collective benefits for city residents and truly democratize the local economy, citizens cannot act alone. As such, our model of co-governance implies the involvement of other actors including public authorities, private enterprises, civil society organizations or NGOs, and knowledge institutions. The only question is how to think about, or conceptualize, their involvement. Building on the idea of the helix from innovation studies, we propose a *quintuple helix* or 5P co-governance model that integrates the literature on innovation ecosystems, engaged universities, and citizen science, participatory or deliberative democracy, and governance of common pool resources. It also argues that communities need learning and digital and financial tools to completely realize the kind of co-governance that we envision, and it offers examples of what those tools do and can look like.

The chapter contrasts co-governance with participatory policies, such as participatory budgeting, and other forms of decentralized local decision making. Even the highest form of participation and citizen power can fall short of altering the unequal power dynamics, privileges, and advantages that often characterize urban geographies that are stratified by class, ethnicity, immigrant status, and race. The challenge for any system of participatory or collective governance is to avoid replicating the very inequalities and power dynamics that they are often set up to address. The best collaborative urban processes, in our view, will intentionally and deeply engage and empower the most vulnerable stakeholders in any *partnership* process, arrangement, or agreements. The chapter offers examples of cities that are experimenting with institutional and organizational public-community partnerships (PCPs) and public-private-community partnerships (PCPPs) that target areas and populations exhibiting poor health, social, and economic outcomes. These partnerships are aimed not only at improving the quality of urban space and infrastructure or strengthening community social ties but also at leveraging constructed urban commons as platforms to generate collaborative economies that provide communities in these neighborhoods with the opportunity to develop new skills, support job creation, and offer childcare and other shared services. These examples also reveal other innovations in designing an environment that is conducive to co-governance arrangements throughout a city, such as the importance of administrative mediators (i.e., the neighborhood architect) and institutional spaces (i.e., the co-labs and the collaboratory) that facilitate public, civic, and private actors to collaborate before institutionalizing the alliance through contractor or legal partnerships. These examples also demonstrate the crucial role that digital and technological infrastructures play in increasing the capacity of vulnerable communities to engage in partnerships with other actors as part of urban co-governance.

Chapter 5 explains the five design principles that characterize a co-city: a city that enables its infrastructure to be utilized as a platform on which a variety of urban actors cooperate and collaborate to govern and steward built, environmental, cultural, and digital goods through contractual or institutionalized public-community partnerships or public-community-private partnerships. These partnerships involve cooperation and collaboration between civic, knowledge, public, and private actors that support the creation and governance of shared and common resources by an identified group of people or community vested with the responsibility of maintaining and keeping accessible (or affordable) the resource for future users and generations. These recurring characteristics, methodologies, and techniques best define the ways in which the city can operate as a cooperative space in which various forms of urban commons can emerge and can be economically, socially, and ecologically sustainable. Some of the design principles described in this chapter resonate with Elinor Ostrom's design principles, whereas others reflect the reality of constructing common resources in the context of contemporary urban environments.

A short concluding chapter, chapter 6, reflects briefly on the challenges that we continue to face in the application of the co-city design principles and pathways for future study and research. The design principles are extracted from the projects that we have surveyed and studied, including some in which we have participated. As the co-city approach has spread to different kinds of cities, we have begun to identify some of the challenges to its application in other political, social, and economic contexts. In this concluding chapter, we identify new challenges from projects in Baton Rouge, Louisiana, and Rome, Italy, which will test the power and saliency of the co-city approach to address endemic racism and injustice in a US city and bureaucratic ossification and wealth concentration in a capital city with one of the richest cultural heritages and most vibrant sustainable innovation ecosystems in the world.

1
RETHINKING THE CITY

Imagine almost any city in the postindustrial US in the 1980s—Detroit, Chicago, New York. Failed urban renewal programs have left most of these places scattered with vacant lots, abandoned by their original owners, and now reclaimed by the city through tax foreclosures. The move of much of the urban population from cities to suburbs is complete. Inner cities are ravaged by a new drug epidemic and escalating crime rates. Now imagine that amid economically and socially fragile communities, neighborhood residents use these vacant lots to construct hundreds of community gardens. Residents sweep away the trash and drug paraphernalia. They plant and cultivate trees, flowers, and vegetables. The gardens become places where residents of different ethnic backgrounds and ages interact, local food is produced, and because the garden participants become the eyes and ears of the community, crime is prevented. The gardens also provide the infrastructure for community interaction—sitting areas with benches and tables, playgrounds, water ponds and fountains, summerhouses—as well as for cultural and social events. These community gardens help to revitalize neighborhoods, once seen as socially and economically fragile, through the self-help of citizens who have transformed these abandoned and underutilized spaces from barren, degraded ones to aesthetically pleasing and productive ones.

Fast forward to the present. Urban revitalization is well under way; many suburbanites who left the city decades ago are now itching to return to the promise of safe, burgeoning city life. Private developers are interested in land once seemingly forgotten. City officials, too, are interested in previously abandoned lots, particularly in selling them to private developers for the construction of new housing and other developments. Toward this end, city officials announce plans to bulldoze hundreds of community gardens and sell off the lots to private developers. Neighborhood residents bring a lawsuit to stop the auctioning of the gardens, to no avail. They argue that the gardens to be auctioned off are predominantly in low-income and ethnic minority neighborhoods and that their destruction would disproportionately deprive those neighborhoods, especially the most vulnerable, of the green space and social and economic resources the gardens provide. In response, city officials characterize the lots as *vacant* which, while legally correct under state law, defies the reality of the transformed land and the value of the gardens to surrounding communities. City officials argue that in the long run the communities where the gardens sit would benefit from the new development and promise to devote some of the newly redeveloped land to affordable housing. In the end, many of the gardens are auctioned off for luxury condominiums and parking lots.

Resident transformation of previously vacant lots into community gardens represents a form of local environmental stewardship. Local environmental stewardship consists of actions taken by individuals, groups, or networks of actors, with various motivations, to protect, care for, or responsibly use valuable or scarce resources in pursuit of environmental and/or social outcomes (Bennett et al. 2018, 3). Stewarding reimagines the relationship between humans and these resources through its commitment to community participation in the restoration practices of these resources that often are proximate to local populations and utilized for their subsistence needs and livelihoods (Barritt 2020, 2–3). The residents' (and others') act of claiming and caretaking of vacant spaces to function as social infrastructure is consistent with these foundational principles of environmental stewardship, particularly in the urban environment (Campbell et al. 2021). Stewardship is also connected to the idea of place making—the "intentional effort into the creation of good public spaces to promote people's well-being" (Murphy et al. 2019, 2).

Transformation of the lots into gardens was also, according to the resident's lawsuit, responsive to environmental justice concerns in low-income and minority communities (Foster 2006). The literature on environmental justice has brought attention to racial, ethnic and class disparities in exposure to environmental hazards and the lack of environmental amenities such as green spaces and fresh food in low-income and minority communities (Cole and Foster, 2001). Underlying these disparities are, for example, pre-existing economic, social, and political inequalities that contribute to the social vulnerability of African Americans and Latinos in the US. Other factors include differences in power and access that can prevent some communities from receiving resources or from participating in crucial planning and other decision-making processes that shape their communities. The environmental justice framing in this dispute links the demand to the gardens and the resources they provide with the imperative of creating socially just *and* ecologically sustainable communities—what Julian Agyeman refers to as "just sustainability" (Agyeman et al. 2003). Socially just and ecologically sustainable communities would provide equitable access to green space, clean air and water, healthy food, affordable good-quality housing, and safe neighborhoods.

In their dispute over the community gardens, to highlight the importance to them of the stewarded resources, residents engage in a rhetorical campaign to situate the gardens as the functional equivalent of parks or *parkland*. Parkland is protected in US law by the public trust doctrine, a legal concept traceable to ancient Roman law (Sax 1970). The public trust doctrine recognizes that certain types of property, particularly valuable resources that are difficult to replace are held by the sovereign in trust for the public and imposes strict limits on the sale, transfer, or use of this property for purposes other than those open and accessible to the public. However, while the doctrine's nineteenth-century origins in the US included the protection of both natural resources and their urban equivalents—namely, city streets, public squares, and roadways—the doctrine has since been considerably narrowed (Kaplan 2012). Most modern courts and legal commentators consider the doctrine to be effectively confined to resources and property having some nexus with *navigable waters*, maintaining strict adherence to the Roman law origins of the doctrine. Unfortunately, the community gardens are neither parkland nor a protected resource under the public trust doctrine and thus could not benefit from legal protection on this basis.

The rhetorical promotion of the gardens as parkland, as more than just another piece of undeveloped land, also reflects the residents' anxiety and fear that given the enthusiasm for redevelopment in the city, they too will be displaced along with the gardens. Not only do they stand to lose the physical resources provided by the gardens but also their community and the social ties that bind them to a place they have known and lived in for many decades. The residents' concern is reminiscent of Jane Jacobs's critique of the urban renewal slum clearance programs of the 1940s and 1950s in the US, which resulted in not only the destruction of physical neighborhoods but also the destruction of the irreplaceable social capital—the networks of residents who build and strengthen working relationships over time through trust and voluntary cooperation—necessary for self-governance of urban neighborhoods (Jacobs 1961).

The rhetorical tension or battle between the residents and the city reflects two competing understandings of the land on which the gardens are located. The city characterizes the land as atomized space—*vacant*—separate from the social fabric of the surrounding community and the human activity that gives the land its value. Severing the resource from its social and economic function to the surrounding community and the value of the interaction space that it creates, the city is able to turn it into a purely commodified asset to be sold on the market to a private developer. In this way, the city positions itself much as a private property owner would. It can use or dispose of the property that it owns as it sees fit. The city's position is reflected in a quotation from an elected official referring to the acres of city-owned vacant land: "when the city owns the property, we get to call the shots about how land is developed and for whom, which is why these properties are so valuable" (Kinney 2016).

Residents, on the other hand, value the land as a collective shared resource rooted in the resource's relationship to the surrounding community and the solidarity borne from the interaction spaces the resource provided. Through the cultivation and stewardship of land, these communities have strengthened their collective capacity and resiliency to survive and thrive, even as their governments pursue urban growth strategies that build up the urban core and downtowns but neglect their neighborhoods.

WHO OWNS THE CITY?

The conflict between residents and the city in the preceding example reflects and highlights the tension between the exchange and the use value of urban land. The tension between exchange and use value, as John Logan and Harvey Molotoch famously argued, most often plays out at the neighborhood level with residents defending the use of land to satisfy the essential needs of everyday life, build informal support networks, establish security and trust, capture agglomeration benefits, and fortify shared identity (Logan and Molotoch 1987, 103). This tension continues today but arguably at a heightened level, in large part because urban land values are at historic highs. The total value of US urban land is estimated to be $25 billion, roughly more than double the nation's overall economic output or GDP (Albouy et al. 2018, 459). Nearly half the total value is packed into just five metro areas: New York, Los Angeles, San Francisco, Washington, DC, and Chicago, with land in and around the urban center being the most valuable. These cities, and their international counterparts, have become what some scholars call "exclusionary megacities" that share a "property-centered approach" to urban growth that "prioritizes the maximization of existing property interests" and "is premised on the drive to maximize the value of land for current owners . . . against the interests of middle- and low-income populations" (Pritchett and Qiao 2018, 474).

The conflict between highly sought-after cities and some of their communities over control of property and resources that are in the public domain brings into view the question increasingly being asked by those who live in and study cities. As sociologist Saskia Sassen provocatively muses, "who owns the city?" in an era of "corporatizing access and control over urban land" that is transforming the "small and/or public" into the "large and private" across so many cities around the world (Sassen 2015). Sassen details a post-2008 pattern in many cities around the world of large corporate entities acquiring "whole blocks of underutilized or dead industrial land for development of high-end luxury commercial and residential space." This large-scale privatization of land in cities, she points out, has an effect (oddly) of "de-urbanizing" city space and creating de facto "gated" spaces with lots of people. Consequently, the scale of spaces that are accessible to the public is shrinking and the population of those displaced from cities is expanding (Sassen 2014). Her question is shared by many others concerned that public officials are commodifying and privatizing our collective resources in cities, disproportionately harming those who lack private resources and who most depend on public resources.

Of course, not all urban land is valuable. This is particularly true in so-called minimal cities like Detroit that either were on the verge of or have declared bankruptcy (Anderson 2014). For these cities, valuing vacant land for its exchange value is most often borne out of economic necessity. These local governments are actively trying to place the land back into productive use, typically by acquiring title to these properties and placing them in a land bank or public receivership until title is clear for their transfer to a private investor and corporate developers. This complicates but does not resolve claims by residents to *share* these resources with communities that have been cultivating and stewarding land and are apt to view this available urban infrastructure as opportunities for collective bottom-up management of their communities.

Consider what has happened in Detroit, a postindustrial city that has experienced serious decline over the last few decades as its workforce in factories began to wither away and white residents fled to the suburbs, leaving a predominantly Black population to struggle for more equitable housing conditions and political power (Boggs 2012). The majority Black city is still struggling to come back from its 2013 bankruptcy. Shortly before Detroit became the largest American city to declare bankruptcy in 2013, the city began redistributing public property, increasing tax foreclosures, privatizing public services, and increasing private investments into the city (Safransky 2017, 1082). In its quest to raise money in a context in which there is no regional tax sharing with its more affluent suburbs, Detroit has been characterized as a *predatory city*—a reference to the claim that public officials are systemically dispossessing predominantly Black residents of their homes through, for example, illegitimate property tax foreclosures (Atuahene 2020).

Today, Detroit is often referred to as a *tale of two cities*. In one of these cities, private capital is fueling development in the Downtown and Midtown areas, including areas close to some of the city's universities and hospitals, populated by gentrifying young white professionals. The other city consists mainly of Black residential neighborhoods populated by long-time residents who have not been able to or have not wanted to flee the city during its darkest days (Alvarez and Samuel 2018).

Detroit's land bank program has been a sore point with many of its long-time residents who are not directly reaping the benefits of the city's downtown and midtown revitalization. The Detroit land bank authority holds the title to the majority of the city's 43,000 vacant homes (some estimates are as much as 68 percent) that it has acquired through tax foreclosures and to thousands of vacant lots. Local newspaper stories recount the frustration of many residents, particularly those living in neighborhoods still in decline, trying to acquire property through the land bank. To some observers, the city has had no problem selling hundreds of parcels of vacant lots to large corporations to expand its commercial urban tree nurseries in Detroit, renovate dilapidated homes, and free up land for a car assembly plant (Livengood 2019). Small-scale, long-established Black farms in the city, in contrast, have had difficulty purchasing the land on which they farm from the land bank. According to one account, despite their interest and attempts to purchase the land on which they have stewarded acres of farm sites that serve the needs of food-insecure homes and neighborhoods, Black farmers have been unsuccessful in convincing the city to allow them to purchase the land (Baker 2020, 28–29).

As Sara Safransky (2017) has documented, drawing on original interviews and observations at public meetings, many Detroit community members and activists are concerned about the "top-down re-territorialization" approaches in the city that do not take into consideration historical attachments to these lands, the people that are most impacted, and how those people are included or excluded from the narrative of urban revitalization. Many Black community members believe that this land is "black man's land" and serves as a site of historical and collective memory. There is an emotional and physical connection to Detroit as a product of the Great Migration much as there is an emotional and physical connection to the rural land of the US southern region that Black families left behind for economic and political advancement (Mitchell 2005). For many of these residents, urban agriculture is a means to an end in the long-running struggle for social justice. As Safransky notes, the claim to resident-stewarded land "is one part of a broader struggle to re-appropriate modes of social reproduction to serve the community rather than capital" and part of grassroots organizing efforts "seeking to undo colonial spatial orders and structures of white supremacy by building new organizing infrastructures, commons-based institutions, decentralized forms of governance, and social and ecological relationships" (Safransky 2017, 1093).

Bottom-up approaches to land vacancy often focus on greening and urban agriculture as a viable use of the land (Bentley et al. 2016). Increasingly, however, residents are asserting their rights to these vacant structures for affordable housing units, community shelters, health facilities, child-care facilities, artistic or entrepreneurial spaces, and in many cities around the world, housing for the homeless (Alexander 2015, 2019). Beyond vacant land, residents, and communities in many cities around the world view abandoned homes, factories, strip malls, and other structures as opportunities for productive reuse. Residents and communities often desire more control over vacant land and structures—available urban infrastructure—to remake spaces and to meet the basic needs of economically and socially vulnerable populations.

In some US cities, for instance, mothers have led the movement to occupy vacant homes, addressing homelessness and housing instability at a time when the COVID-19 pandemic made paying rent unmanageable for many and despite the risk that they could be removed at any moment by the state or local government. In Philadelphia, over forty people, mostly single mothers, occupied boarded-up vacant homes owned by the Philadelphia Housing Authority (PHA), the largest landlord in Pennsylvania. A coalition of groups such as Occupy PHA, Black and Brown Workers Cooperative, and the Revolutionary Workers Collective argued that the PHA had become indistinguishable from a private developer and planned to let them sit idle until they found an interested buyer, leading to gentrification and displacement while ignoring the needs of the city's low-income residents (Tribone 2020). The coalition argued that instead of being sold to developers who would build market-rate housing, the homes should be transferred to a community land trust that would repair and manage them in perpetuity as affordable housing for the city's poorest residents (Phillips 2020).

In orchestrating the occupation, advocates were inspired by the actions of Moms4Housing, a collective of Black mothers who are homeless and marginally housed in Oakland, California. Moms4Housing occupied an empty house owned by a real estate company, and they were evicted. Fortunately, a few weeks after the eviction, the real estate company agreed to sell the house to the Oakland Community Land Trust, which acquires land for the benefit of low-income communities (NPR 2020). A similar group of mothers in Los Angeles took inspiration from the Oakland collective in March. Under the name Reclaiming Our Homes, the housing-insecure Oakland families began to move into vacant homes owned by Caltrans, the state transportation authority. They called themselves reclaimers of the property and argued that it was unacceptable that usable homes owned by the state were lying empty when people were homeless and living on the street. None of the reclaimers were evicted and state officials agreed to lease more than twenty of the houses to the city's housing authority, which then allowed a dozen families to live in them for two years, part of a transitional housing program (NPR 2021).

THE CITY AS A COMMONS

The idea that urban land and infrastructure are more than assets for exchange on the real estate market and are more akin to a common resource that can and should be shared with urban residents brings into view the argument and ideas developed in this book. The efforts of communities to access and utilize vacant or underutilized property and other public resources, particularly in structurally disadvantaged communities, have the potential to capture positive value to create goods (both tangible and intangible) that can be shared and stewarded by these communities. The claim to available urban infrastructure, particularly that is in the public domain or under the control of the state, does not rest necessarily on the desire for *ownership* of land or a desire to exploit its exchange value. Rather, it is based on recognition that the built environment constitutes a variety of potentially shareable and stewarded urban goods that can generate essential resources for urban residents lacking those resources.

In thinking about cities and their resources in terms of the potential for shared stewardship by urban communities, we have found inspiration in the Nobel Prize–winning work of Elinor Ostrom (Ostrom 1990). In her groundbreaking work, Ostrom overturned decades of economic thought that suggested that there were only two ways to manage and govern shared resources: public control or private ownership. Ostrom found examples all over the world of resource users cooperatively managing and stewarding a range of natural resources—uncultivated lands, fisheries, communal forests, groundwater basins, and irrigation systems—using "rich mixtures of public and private instrumentalities" (Ostrom 1990, 182). In Ostrom's examples, resource users devise and enforce their own rules for sustainably using and sharing the resource without overconsuming or depleting it. Importantly, these rules and the community's right to enforce them were recognized by external governing bodies and public agencies.

Ostrom's work explicitly refuted the assumption, most famously attributed to Garret Hardin in his classic essay "The Tragedy of the Commons," that without exclusion rights individual users could not overcome collective action problems and work together to manage resources that were open or shared. Hardin's stylized tale of tragedy unfolds in the context of a "pasture open to all" on which each herdsman is motivated by self-interest to continue adding cattle until the combined actions of all the herdsman results in overgrazing, eventually depleting the resource for everyone (Hardin 1968). Unlimited access to shared resources inevitably leads to overconsumption and complete destruction of the resource. As Hardin argued, absent a system of state management or governance it would be difficult, if not impossible, to restrain the impulse of users to pursue their individual self-interests even when pursuit of those interests results in the degradation or exhaustion of the resource. Hardin concluded that such "freedom in the commons"—that is, the lack of controls on individual behavior and self-interest—"brings ruin to all." Ostrom rejected the public/private binary choice of solutions that Hardin offered to avoid the tragedy, successfully demonstrating that the choice between central government regulation and private property rights does not capture the full range of approaches to managing or governing the commons.

Looking at shared urban infrastructure through the lens of the commons is one way to acknowledge the potential of cities to be generative and its resources shared among communities of users under certain circumstances. Our conception of the urban commons builds on Ostrom's insights and her methodology but is adapted to urban environments on the basis of our observations of the dynamics and challenges specific to those environments. However, as we discuss in chapter 2, there are important differences between the collectively governed and stewarded natural resources that Ostrom studied and the *constructed* urban resources that are the subject of our research. One key difference that we highlight is the presence of a strong *enabling* state (the local government or public authorities) and *pooling economies* that bring together other actors and resources to support the co-production and co-governance of urban commons. These resources can be stewarded by allocating rights and responsibilities in a way that gives communities decision-making use and control over them in a manner similar to ownership and vests them with the duty of maintaining and keeping accessible (or affordable) the resource for future users and generations. As we discuss in this chapter and the next, institutional arrangements such as community land trusts and limited equity cooperatives, among others, can be vehicles for property and resources stewardship consistent with the idea of the urban commons.

Cities and much of their infrastructure share some of the classic problems of what Ostrom and other economists refer to as a "common pool resource"—an economic term that signifies the difficulty of excluding people from a resource, which leads to rivalry for its resources and the need to design effective rules, norms, and institutions for resource management and governance. Today, as has been true in the past, all kinds of people flock to cities to create and to re-create their lives by accessing and exploiting the physical, social, and cultural resources that are uniquely found in dense and diverse urban environments. It is the very openness of cities and many of their resources that makes them intensely rivalrous—subject to competition for land and other resources. As law and economics scholar Lee Fennell posits, "the city analog to placing an additional cow on the commons is the decision to locate one's firm or household, along with the privately-owned structure that contains it, in a particular position within an urban area" (Fennell 2015, 1382). Cities represent a particularly challenging collective action problem, Fennell argues, in figuring out how to achieve the benefits of proximity among people and land uses while curbing the negative impacts of that same proximity. As a consequence of the rate of urbanization around the world, cities have a "participant assembly" problem that requires finding the right trade-off between the positive spillovers of agglomeration or proximity and the negative impacts of congestion (Fennell 2015).

Hardin's "tragic" tale can easily be told about cities and the different kind of shared resources within them. Because of the difficulty of excluding people, a city can easily become heavily congested, its resources strained and eventually diminished. City streets, urban parks, cultural resources, vacant land, and even neighborhoods can mimic the kind of "tragedy of the commons" resulting from the self-regarding actions of others that lead to the degradation or destruction of the resource. The tragedy of the (urban) commons arguably sealed the fate of many American cities in the 1970s and 1980s due to the lack of sufficient management and governance of shared urban resources amid declining public resources to properly care for them. Roy Rosenzweig and Elizabeth Blackmar's history of New York City's Central Park, for example, recounts how, after years of opening the park to permit a wide variety of events and groups to use the park, Central Park quickly became a space in which access to the "whole community" posed inevitable conflicts and competition between users (Rosenzweig and Blackmar 1992). Many saw the park as deteriorating rapidly due to its openness to various events and a potpourri of users, resulting in increased maintenance and cleanup costs that the city was not able to absorb. This deterioration escalated with the onset of the fiscal crisis in the 1970s and the decline in city appropriations, which devastated the entire urban park system, leaving many parks and recreational areas unsafe, dirty, prone to criminal activity, and virtually abandoned by most users.

The "tragedy" of urban parks thus unfolded during a period of "regulatory slippage," which is a significant decline in the level of local government control or oversight of the resource, for whatever reason (Foster 2011). When local governments abdicate control or stewardship over common resources, often due to declining tax bases and limited resources, these resources can resemble less a public good (nonexcludable and *nonrivalrous*) and more a traditional common pool resource (nonexcludable and *rivalrous*). Without proper stewardship of the resource by central authorities these resources become contested or rivalrous through the competing demands and uses from a variety of actors, whether they are ordinary pedestrians, opportunistic criminals, or frequent park users. Such users might be tempted to use or consume the common resource in ways that rival and/or degrade the value or attractiveness of the resource for other types of users and uses. Competing user consumption and demand may overwhelm or confound the ability of government to manage this competition and lacking such management, the increase in certain types of uses of common space such as excessive loitering, aggressive panhandling, graffiti, or littering will eventually begin to rival other users and uses of this space.

A similar "tragic" story can be told about the *neighborhood commons* in many urban communities. The quality of a neighborhood commons—of its street life, culture, sidewalks, open or vacant spaces, and public parks—might begin to decline through increasing and competing demands by different users and uses of the space. This kind of urban tragedy can result from an increase in what Robert Ellickson (1996) called "chronic street nuisances"—excessive loitering, aggressive panhandling, graffiti, or littering—that eventually begin to rival, if not overwhelm, other users and uses of open spaces. Overuse or unrestrained competition in use of the space creates conditions that begin to mimic the type of commons problem that Hardin wrote about—that is, such resources become prone to degradation and decline. On the other hand, deterioration in the quality of a residential neighborhood might occur because of the failure of property owners to make repairs that would be economically rational only if other owners took steps to improve their structures as well (Oakerson and Clifton 2017, 416). In some cases, we might view neighborhood decline as the consequence of a failure of residents to collectively govern their common pool resources, perhaps because of a lack of adequate capacity among residents to self-organize or a failure to sustain mutual assurance because of holdouts, for example, homeowners who refuse to cooperate, allowing the neighborhood to deteriorate regardless of what their neighbors do (Oakerson and Clifton 2017, 422).

The rise of many place-based governance institutions like business improvement districts (BIDs) and park conservancies are often in response to the "tragedy" of the urban commons—the general decline of city streets, parks, and commercial areas under conditions of regulatory slippage, which call into question the ability of public administrations to steward these collective resources. Park conservancies, BIDs, and other public-private partnerships in the 1980s and 1990s in US cities are widely credited for the revitalization of urban parks, streets, and commercial areas at a time when those resources were suffering from public neglect and during times of fiscal strain on local governments. These sublocal entities work with local park agencies, police departments, and other city officials to enable private property owners and other local interests to manage and govern shared community resources with a high degree of operational autonomy. We discuss these institutions in somewhat more detail in chapter 2. For now, we note only that in some ways they resemble the kind of "nested" enterprises that Ostrom found in the natural environment, in which resource users work with government officials, agencies, and other stakeholders to collectively manage and govern a shared resource. In other ways they resemble a privatized city in which the responsibility for public spaces is placed in the hands of corporate and commercial interests that increasingly invest in and manage common shared resources like parks and street-level amenities.

While the long-term sustainability of these place-based governance institutions is often a virtue in an environment of "regulatory slippage," we must be mindful of the dark side of these institutions—the risk that shared resources can be captured, co-opted, and enclosed in ways that undercut the very public nature of the resource. Some enclosure and even exclusion of others from a resource might be necessary to maintain it and keep it accessible to some populations and communities. But how much, by whom, and toward what ends are questions that require us to scrutinize institutions, such as some BIDs and park conservancies, that manage large-scale public resources or commons for heterogeneous users. In some circumstances, institutionalized forms of small-scale user governance can create new social problems and divisions, particularly when "insider" group norms designed to maximize group welfare do so at the expense or exclusion of nongroup members (Ellickson 1991, 169). These collectively governed institutions created to solve social and economic problems can become what Brigham Daniels has called "tragic institutions" that develop or expand their "grip" on common resources in antidemocratic or exclusionary ways (Daniels 2007).

It is not enough, in other words, to enable new forms of collectively governed institutions throughout a city that manage public or shared resources. We must also be able to assess whose interests are served by those institutions and how easily they can be captured and their value extracted in ways that aggravate and deepen social and economic inequality. As

David Harvey has noted, many different social groups can engage in local self-governance of shared resources for many different reasons—"the ultra-rich, after all, are just as fiercely protective of their residential commons as anyone, and have far more fire-power and influence in creating and protecting them" (Harvey 2012, 74). Harvey offers the example of a new or revitalized urban park, such as New York City's famous High Line, which is easily capitalized upon by surrounding property owners who capture the value of the common good through drastically increasing surrounding property values and the extraction of rents. The result is to make unaffordable and inaccessible for most of the city the housing that surrounds the park. Thus, although "open and accessible" to all, some kinds of new urban commons can have an exclusionary effect, "radically diminish[ing] rather than enhanc[ing] the potentiality of commoning for all but the very rich" (Harvey 2012, 75).

CONSTRUCTING URBAN COMMONS

The types of collectively governed enterprises we refer to as *urban commons* emerge less from the "tragedy of the commons" and the need to simply create another layer of institutional management for public or shared resources. Instead our embrace of the commons as a framework for urban resources stewardship resonates with the idea of a "constructed commons," which is the result of emergent social processes between resource users, communities, and other stakeholders (Madison et al. 2014). Constructed commons grow not out of the "tragedy" of shared common resources but rather out of what legal scholar Carol Rose (1986) refers to as the "comedy of the commons." The comedy of the commons involves granting access to resources that the community values and that increases the solidarity between urban residents. Rose found that some British courts considered as "inherently public property" resources such as land, open space, and roads that were customarily used by the public for gatherings or other activities valued by the community. These were activities in which "increasing participation enhances the value of the activity rather than diminishing it" (Rose 1986, 768). Instead of tragedy or overconsumption in these spaces, Rose argued that we are more likely to find "comedy"—the "more the merrier." The more that people come together to interact, Rose observes, the more they "reinforce the solidarity and well-being of the whole community" (Rose 1986, 767–768).

Under certain circumstances urban resources are sufficiently available in supply to present a "comic" scenario in which open access or shared resources are less the site of a potential "tragedy" and more a platform that enables actors to enjoy and produce reciprocal positive spillovers that generate increasing returns to scale. On this theory, it is urban "interaction space" that renders many public spaces and resources so valuable (Fennell 2015). In other words, interaction space facilitates a host of other goods—knowledge exchange, social capital accumulation, solidarity, access to material resources such as recreation space or food—that accrue to individuals in close proximity to one another. Capturing the positive gains of urban "interaction space" occurs in collectively created, produced, and stewarded resources like community gardens, as in the opening example, or in the construction of informal settlements on the urban periphery of many major cities in the Global South, referenced later in this chapter.

Beyond the positive spillover effects that shared urban spaces and infrastructure can have for social interaction and solidarity, we are chiefly interested in the ways that these spaces and resources can be used to construct new resources and services for disadvantaged populations and communities. As such, we define constructed urban commons as those that result from a process of bringing together a spectrum of actors that work together to co-design and co-produce shared, common goods and services at different scales from existing shared urban infrastructure.

In US cities, for example, vacant or abandoned land and structures is more ubiquitous than most people realize. This is true whether we are talking about resurgent or declining cities, or their suburbs. Vacant land constitutes anywhere from 15 to 20 percent of older so-called *legacy* cities such as Philadelphia and Detroit and consists of thousands of vacant lots which are often concentrated together in a pattern of "hypervacancy" (Mallach 2018). There are more than 120,000 vacant lots in Detroit—nearly forty square miles, a third of the city. Philadelphia has an estimated forty thousand vacant lots with no known use. Even so-called *magnet* cities like New York, Seattle, Los Angeles, San Francisco, and Washington, DC, have their fair share of vacancy, with rates ranging from 5 to 15 percent, even as these cities experience unprecedented growth (Mallach 2018).

Vacant land and structures are often concentrated in neighborhoods that suffer from a history of neglect and underinvestment, creating both an opportunity and a risk for residents that undertake to construct new urban resources or goods that may not be sustainable given their precarious claim to the resource. As studies have shown, high vacancy rates are positively associated with displacement and gentrification (Morckel et al. 2013), a risk that is particularly strong in neighborhoods with clustered residential and commercial vacancies because they attract new investors and catalyze redevelopment (Lee and Newman 2021). For this reason, residents often desire control over this land to rebuild their communities and to stave off the threat of displacement and gentrification. They also desire to utilize vacant land and structures, as indicated previously, to provide critical goods and resources and to cure the social and environmental injustices that these communities have lived with over decades.

Community land trusts (CLTs) are one kind of constructed commons that are flourishing around the world as a vehicle to allow community control of land toward stabilizing communities vulnerable to being displaced by market forces. CLTs are often used to acquire and develop available urban land and structures to create affordable housing, commercial space, and green and recreational resources in urban communities that lack those resources. CLTs are emerging in cities and urban communities all over the world as a form of land stewardship to preserve housing and other land uses as affordable and accessible for future generations and to promote development without displacement (Davis 2010). Local public authorities often facilitate these resident and community governed institutions by making available vacant urban land and structures, and by expending public dollars to subsidize these arrangements.

Consider what occurred in the Dudley Street neighborhood of Boston in the late 1980s and early 1990s, which was known at the time as one of the poorest areas of Boston. Neighborhood residents worked with city and state officials to acquire and utilize over fifteen acres of city-owned vacant lots and fifteen acres of privately owned, tax-defaulted vacant lots to revitalize Nubian Square (formerly Dudley Square). Residents and various community-based institutions planned to create an "urban village" consisting of affordable housing, urban farms, community gardens, and other neighborhood amenities (Medoff and Sklar 1994). Dudley Street

Neighbors Initiative (DSNI), the nonprofit formed to oversee the process of creating the urban village, set up a community land trust into which the over one thousand parcels of land were placed. The land trust controls over thirty acres of land and is currently trying to acquire additional land (Smith and Hernandez 2020, 288). We return to the Dudley Street example in chapter 2.

Community land trusts also are emerging in informal settlements in Latin America that are facing rising land values and impending gentrification in neighborhoods once considered to lack significant market value. Recently, a collection of eight neighborhoods in San Juan, Puerto Rico, became the first informal settlement use the community land trust as a response to this challenge (Algoed et al. 2018). The CLT was founded to preserve and develop informal neighborhoods along the Caño Martín Peña Canal and to protect them from involuntary displacement and gentrification in the now collectively controlled over 270 acres of land that previously belonged to government agencies. The CLT worked jointly with NGOs, universities, foundations, the municipality of San Juan, Puerto Rican public agencies, and US federal agencies to dredge the nearby river. The land is now owned and managed by the CLT, which as of 2019 included approximately fifteen hundred low-to-moderate-income households, whose purpose is to ensure the availability of permanently affordable housing and "serve as an instrument for the generation and redistribution of wealth" (Hernández-Torrales et al. 2020). There is a similar effort to place favelas, or informal settlements, on the outskirts of Rio de Janeiro, Brazil, in a CLT as a response to threatened gentrification and land insecurity in these well-established communities with long histories of cultural production and community investment (Williamson 2018).

As we describe in the next chapter, the use of community land trusts and other limited equity mechanisms for holding and governing urban land and infrastructure is a way of keeping these resources accessible and affordable to a broad range of users while allowing communities of users to steward these resources over time. They do so by taking these resources off the speculative market and separating land *ownership* from land *use* while creating the possibility for users, such as housing occupants, to *sell* their interests back to the trust or cooperative for limited equity. Land trusts and limited equity cooperatives are examples of constructed urban commons because they are collectively created and governed by their users, and they are decommodified to the extent that keeps them accessible and affordable to communities vulnerable to expulsion from urban land markets (Huron 2018). Although they are similar to other kinds of shared, collectively governed property such as condominiums and traditional co-ops, their purpose is distinct—namely, to remove the profit motive from land use. Instead of being valued for their potential exchange value on the market, the land is valued and operated according to its use value to the surrounding community (Huron 2018, 7).

In addition to CLTs, our work and that of others has identified numerous other examples of constructed urban commons that result from residents working together with other public and private actors to generate new forms of common goods using the existing infrastructure of the city. The examples include wireless mesh or broadband networks, energy microgrids, and other essential social infrastructure. Community-created and user-managed *mesh* networks, for example, are decentralized wireless access points connected to each other in a defined geographic area (De Filippi and Tréguer 2015). These networks have been established in many European and US cities, utilizing existing urban infrastructure and the combined efforts of many local actors—including public and private property owners who grant access to buildings and other structures to mount the access points—to create a solution to the *last-mile* connectivity gap. The goal of these networks is to bring internet service to communities and populations that lack high-speed wireless or broadband access, increasingly seen as a necessary public good or *fourth utility*. In this model, no one owns the entire infrastructure (open and free access), but everyone who wants access can contribute with their own resources to run the network, which is managed and governed by the community. The *digital stewardship* in places like Red Hook, Berlin, and Detroit are often grounded in *digital justice* principles including equal access, participation by historically excluded populations, common ownership through cooperative business models or municipal ownership, and healthy communities that promote economic development from within and expand educational opportunities and environmental justice (Detroit Digital Justice Coalition n.d.).

We explore many examples of these kinds of constructed *urban commons* throughout the book. Some of these collectively produced and governed efforts are longstanding enterprises or institutions dating back many decades. Others are fairly new and experimental. Some face predictable risks and challenges in sustaining themselves in thick urban environments.

URBAN POOLING ECONOMIES

The social, legal, and political complexity of many contemporary cities makes it challenging for communities, especially those with fewer resources, to steward land and other urban resources. In part, the reason is that it is difficult to completely sidestep the market and state actors. In other words, communities cannot operate as completely independent authorities over urban land or infrastructure. Local governments often have proprietary *and/or* regulatory authority over their infrastructure including vacated or abandoned land and structures in the public domain. As such, constructing urban commons and sustaining these efforts in most cases require an important state role. Local and provincial government actors often need to aid and form a solid alliance with communities to advance collective governance of shared urban resources as well as to scale those efforts across a city.

Consider the construction of informal settlements in and on the periphery of many cities in the Global South. In a process that Teresa P. R. Caldeira calls "peripheral urbanization," new migrants utilize urban land to build their own homes and their communities step by step "according to the resources they are able to put together at each moment" (Caldeira 2017, 5). Caldeira argues that in places such as Sao Paulo, Istanbul, Mexico City, and Santiago, urban residents are agents and not just consumers of urban spaces developed and regulated by others, claiming the city as their own. In a complex process involving many actors, she describes the creation and development of these peripheral urban areas as involving several layers of improvisation and irregularity, as well as negotiation between many actors and agents involved in the process. Although residents are the main agents of the production of the current space in peripheral urban areas, the state is also present in numerous ways as it "regulates, legislates, writes plans, provides infrastructure, policies, and upgrades spaces," frequently after these spaces have already been built and inhabited (Caldeira 2017, 7).

The state role in the bottom-up management of urban land and other infrastructure can become complicated when local governments are forced to compete with or heavily rely on mobile capital to finance development activities. State actors may end up being more accountable to developers or mobile capital than to local communities. At the same time, it is a challenge for disadvantaged communities to create urban commons that can be stewarded in ways that are not co-opted by the market and yet are able to leverage private actors and their resources. In contexts in which the state is weak because of either corruption or lack of resources, communities often need to collaborate and manage resources with other actors such as knowledge institutions, civil society organizations, and the private sector. In some situations, market actors may emerge as the most feasible means to enable the pooling of human, economic, cognitive, and other kinds of resources needed for collective action and collaborative management of urban resources, particularly constructed urban resources. The market could subsidize the commons if proper legal structures, participatory processes, and accountability measures were put in place and there were sufficient social and political capital among resource users to negotiate with market actors.

However, because these conditions are often lacking and communities are often no match for powerful political and economic interests, stewarding urban resources requires pooling the efforts of many urban actors or sectors with the state operating in a supportive and facilitative role. The co-production of goods and sharing practices are spreading in cities all over the world through what we refer to as *pooling economies* (Iaione and De Nictolis 2017). Pooling human capital, efforts, and resources is a crucial feature of the networked economy and the commons, as scholars such as Yochai Benkler have argued (Benkler and Nissenbaum 2006; Benkler 2016). Pooling economies foster peer-to-peer approaches involving users in the design and production process of constructing common goods and services. Peer-to-peer or user-to-user initiatives result in enterprises that are collectively owned or managed and democratically governed and do not extract value out of local economies but rather anchor jobs, cultivate respect for human dignity, and offer new forms of social security. The concept of a pooling economy is closer to what others have referred to as the *solidarity economy* or *collaborative economy*. As such, we use the concept of pooling to capture the collective creation of new kinds of goods and new economies in the city that are distinct from *sharing* economies that rest too often on the commodification of shared goods. Pooling economies are often catalyzed and fostered by community-led initiatives in which residents and resource users partner with various actors to share expertise and resources.

Public authorities play an important role in facilitating a variety of these distributed and co-governed enterprises within a city, enabling urban inhabitants to actively take part in the regeneration of their neighborhoods, create shared goods to sustain themselves and flourish, and develop and nurture the communities to which they belong. State actors—centralized or higher-level authorities—can support a variety of distributed, co-governance enterprises and urban commons throughout their territory. One of the ways that central governments can support these efforts is to reduce the costs of cooperation and help relevant actors to leverage their efforts to achieve high economic and social payoffs from their collective action. This support might include regulatory changes and fiscal or technical support that remove barriers to cooperation or make it more beneficial or convenient for individuals to engage in cooperative behavior.

Whether through regulatory "nudges" like the collaboration pacts adopted by some cities, discussed in the introduction and in chapter 3, or the transfer of financing or physical resources to community land trusts, local authorities are incentivizing and increasing the capacity of communities and other stakeholders to engage in co-design activities and co-governance projects. One recent example is the effort to catalyze the creation of community land trusts in major cities through legislation and public financial support. For example, in 2017, New York City passed legislation that allows the city's Housing and Preservation and Development agency (HPD) to enter into agreements with community land trusts where the CLT is a recipient of city funding, property or a tax exemption (N.Y.C. Admin. Code § 26-2001). The city also appointed a director of CLT initiatives at the HPD, the city's housing agency. Subsequently, HPD allocated a $1.65 million grant to support the development of community land trusts around the city as well as a learning collaborative to build the capacity of nascent CLTs. An additional $870 million was allocated in

2019 to incubate and expand CLTs to develop permanently affordable housing and curb displacement in low-income NYC neighborhoods.

Another way that local authorities enable resource pooling and the creation of urban commons throughout a city is by providing settings and institutional platforms like city-based urban labs. Urban collaborative labs or urban living labs bring together urban stakeholders to participate in the co-design and co-construction of solutions to neighborhood or city-based challenges and more generally to experiment, innovate, and scale those solutions (Chronéer et al. 2019). Mexico City's Lab for the City (Laboratorio Para La Ciudad) is one standout among the many city labs that have emerged over the last decade. The Lab was created in 2013 at the request of the newly elected mayor of Mexico City (CDMX) and operated until its dissolution at the end of the mayor's tenure in 2018. It was led by a young, multidisciplinary team, most of whom had no prior governmental experience, who wanted to abandon a top-down approach to urban governance and orient the new administration toward co-creation of the cityscape. The Lab was designed to be the space where residents, civil society organizations, nonprofits, knowledge institutions, the private sector, and other government departments could pool ideas and resources to realize neighborhood plans that could be capitalized through governmental programs such as the Participatory Budget and the Neighborhood Improvement programs. The Lab's projects tackled a range of issues including sustainable mobility, pollution, public security, road safety, and revitalization of public spaces, especially those that were the most blighted.

In a similar fashion, the NYCx Co-Lab initiative was designed to create a more distributed series of *neighborhood innovation labs* in underserved neighborhoods throughout New York City to leverage smart city technologies to co-design and co-develop impactful technologies with residents in those communities. Funded by the Mayor's Fund to Advance New York City, a nonprofit organization that facilitates public-private (and community) partnerships throughout New York City, the co-labs emerged out of the Mayor's Office of the Chief Technology Officer (MOCTO). Community members are expected to identify the most pressing problems and define the potential solutions that will help historically disenfranchised communities to keep apace as the city's economic and technological prospects evolve. In each co-lab, residents and community-based organization are expected to work alongside civic technologists, startups, tech industry leaders, and city agencies to ensure that the most vulnerable ethnic and low-income communities are placed at the center of the development of a smart city. The outputs of the various co-labs would propose and test new solutions to modernize public infrastructure, support community-driven development, and bridge the digital divide in low-income and ethnic minority areas of the city. The lab's goals also include the creation of pathways for individuals from low-income neighborhoods to become civic leaders through engagement with the technology and the innovation economy that is rapidly changing their communities.

In theory, these labs are designed to nudge urban governance away from a neoliberal model of development that is overly reliant on market-based solutions that privatize public services and commodify urban spaces (Cole et al. 2018). In practice, they are more often entry points into the co-production of a variety of urban goods and resources, mediating the relationship between district- or neighborhood-level institutions and community residents. In this way they are part of an ecosystem of distributed, polycentric systems of decision making within a city or metropolitan area. We discuss urban labs in more depth in chapter 4.

ENABLING A POLYCENTRIC SYSTEM OF URBAN CO-GOVERNANCE

The idea of the state as a facilitator of pooling economies and collective resource stewardship is part of the move from a centralized system of *government* to a system of urban *governance* that redistributes decision-making power and influence away from the center and toward independent and autonomous self-organized units of resource management. Elinor Ostrom (1990) and others referred to this kind of distributed ecosystem of autonomous governance units as *polycentric* to capture the idea that although higher-level governments or officials might take the lead on a large-scale problem, the care and responsibility for shared goods can operate at different levels. Although the central government authority remains an essential player in facilitating, supporting, and even supplying the necessary tools to govern shared resources, in polycentric systems multiple governing entities or authorities operate at different scales with a high degree of independence to make norms and rules within their own domains (Ostrom 2010a). Polycentric systems can unlock what Ostrom called "public entrepreneurship"—opening the public sector to innovation in providing, producing, and encouraging the co-production of essential goods and services at the local level without privatizing those goods (Ostrom 2005b).

Polycentricity is a response to the critique of *monocentric*, top-down governments that exercise monopolies over authority and decision making in complex resource environments. This top-down style of governing is in many cases less efficient and less democratic in the sense that it denies "opportunities for regular citizens to engage in local problem-solving" (Ostrom 2014). Ostrom, her husband, political scientist Vincent Ostrom, and others found that polycentric governance systems are not only capable of successfully and efficiently governing but in many cases are *better* at performing their governance objective than other more centralized, less fragmented governance systems (Gibson et al., 234). In her early study of polycentric police units serving US metropolitan areas in the 1970s, for example, Ostrom concluded that small police units could be just as efficient and often even more so than larger police units (Ostrom et al. 1973). In these early studies, the only actors participating the polycentric systems under review were government actors; non-state actors, such as NGOs, private businesses, community groups, or other actors typically associated with multilevel systems were not featured. Instead, these studies on polycentricism more closely examined questions surrounding the scale of government, and specifically, the most efficient way of organizing *governmental* actors.

Ostrom's later study of common-pool resources, which revealed that individuals are capable of self-governing and working cooperatively in the absence of state or private control, has blinded many to some of the nuances in her work, one of which involves highlighting (not rejecting) the important role played by the state in the creation and maintenance of these self-organized governance units (Ostrom 1990, 133–142). The role of the state is particularly important in complex political, social, and regulatory environments and in the case of large-scale resources. In these settings autonomous units of resource governance, including those that are self-organized, can be "nested" within higher-level state structures. Participants in complex resource systems can benefit from being part of overlapping, nested organizational arrangements. One of the benefits of such arrangements is that they are creatures of experimentalism, adaptation, continual renewal, and "ceaseless innovation" (McGinnis 2016, 10). Although not unique to polycentric systems, the ability to self-correct as they go is one of their often-cited advantages (Ostrom 1998). The ability to self-correct is inextricably linked, at least in part, to another advantage, the "freedom of entry and exit" (Aligica and Tarko 2012, 246). The ability for individuals and ideas to freely flow in and out of the system ensures a constant flow of new updated knowledge, which helps push these complex systems to improve.

But perhaps the key takeaway lesson from a robust body of research on polycentricity is that, at least for large and complex resource systems, "higher levels of state action or support are often necessary to make the lower levels work well" (Mansbridge 2014, 10). An illuminating study by Sarker (2013) on polycentric water irrigation systems in Japan found that whereas operational autonomy is required for the effective operation of a polycentric system, so too is the "financial, technological, statutory, and political support" of the state. Sarker characterizes properly formed polycentric systems as "state-reinforced self-governance" systems, a phrase that captures their need for operational autonomy (self-governance) *and* their equally critical need for state support (state-reinforcement) (Sarker 2013, 739). Fung (2004) came to a similar conclusion in a study of local school districts in Chicago: while the local participants in an education-related polycentric governance "devised the specific means for cooperation and the details of implementation . . . the state at the higher city level provided support, monitoring, and sanctioning for defection" as well as "information sharing across the several local sites" (Mansbridge 2014, 9).

Nevertheless, there are reasons to be critical or cautious about offering polycentric governance resource regimes, even as a partial answer to rising inequality of resources in cities today. The dangers of decentralization are certainly present—the capture of smaller units by economic elites, or the enclosure or privatization of public goods and spaces. Other shortcomings include transaction costs, temptation for free-riding, and the possibility that the coordination effort required for their upkeep will outweigh their potential benefits (Carlisle and Gruby 2019). Most salient for our purposes are questions of power, social-economic conditions that constrain certain participants, and the inclusiveness and fairness of some institutional arrangements. As Gustavo García-López's work has warned, polycentric systems must be attentive to the possibility that key actors are often omitted from collaborative arrangements in which powerful actors tend to prevail and that outcomes are often unequally distributed in ways that reproduce existing power inequalities and injustices (García-López and Antinori 2017; Tormos-Aponte and García-López 2018).

Polycentric structures that are embedded in bottom-up initiatives, researchers observe, present opportunities for more robust participation from historically underrepresented groups and can facilitate experimentation and innovation while overcoming institutional "blockages" at different levels of governance (Tormos-Aponte and García-López 2018). This suggests that it may matter from where and how polycentric systems emerge—from the bottom up, at the community level, or from the top down, a result of state-created institutional structures. We explore this tension between top-down and bottom-up state-facilitated governance arrangements in chapters 2 and 3. Nevertheless, there is a range of ways that the state can be supportive of a polycentric system of co-governance while also checking the opportunistic and exclusionary behavior of individuals and groups, providing a backstop to the failure of internal conflict resolution, helping to monitor compliance with democratic values, and sanctioning noncompliance (Mansbridge 2014, 138–139). This is true even if there are no guarantees in any system against disparities in power, social conflicts, and the unequal distribution of resources that can frustrate even the most well-designed polycentric system.

In short, the state's role in facilitating smaller units of resource governance is critical to realizing the benefits that can flow from a well-constructed polycentric governance system: adaptive flexibility, institutional fit, overcoming coordination problems, building social capital and trust necessary for cooperation and collaboration, and maintaining fairness and equity (Araral and Hartley 2013; Baer and Feiock 2005; Carlisle and Gruby 2019.) At the same time, the complexity and diversity of many urban environments can also magnify the risks inherent in bringing together actors very differently situated in these environments. Some have argued that this kind of system of distributed co-governance simply creates a *third sector* of both informal and formal organizations (or collections of individuals) outside of the state or market, shifting the onus on already vulnerable and overburdened communities to provide for the well-being of urban residents.

We do not equate polycentric governance with devolution of responsibility by the state to provide for the basic welfare of city residents. Rather, the distributed urban co-governance system that we envision and embrace is intended to share the resources of the city to enable communities, particularly those with few resources, to steward common goods responsive to their needs. This sharing requires the state to invest in its neighborhoods and communities as productive units of inclusive social and economic development. The city-based policies described in the following chapters reflect the ways that local governments are moving away from a top-down-oriented resource governance system in which the state monopolistically controls urban resources to a horizontally organized one in which autonomous and collaborative resource governance arrangements become nested within the institutional framework of the city. These initiatives, as we demonstrate throughout the book, can scale with the support of local policies and public resources to create a polycentric network of constructed urban commons in the city.

2

THE URBAN COMMONS

Our approach to the study of the urban commons began with the same question that Elinor Ostrom asked in her groundbreaking studies of natural resource commons (Ostrom 1990): are there groups of residents and/or resource users who are willing and able to organize themselves, work together to establish rules for sharing resources, and monitor themselves in the absence of an external authority or externally imposed regulations? From our observations and empirical research drawing from case studies in over two hundred cities around the world, the clear answer is yes. From community gardens to mesh wireless networks, there are plenty of examples of self-organized groups of users and residents that collectively or collaboratively construct and then manage shared resources. Some of these *urban commons* even share many of the features of Ostrom's natural resources commons governance scenarios. They are often small-scale resources such as a vacant lot, an empty building, a neighborhood park, or wireless infrastructure that rely on the self-organizing efforts of resource users who establish rules of access, use, and distribution of goods or services produced by the resource. Some are larger-scale, more complex resources such as a neighborhood, a large urban park, an urban village, or a large broadband network that local users or communities must work with other public and private actors to construct and collaboratively govern.

Yet, although some urban commons share much in common with the natural resources that were the subject of Ostrom's work, there are facets of collectively governed urban resources that are notably distinct. In this chapter, we tease out and illustrate some of the key distinctions. We first offer a basic introduction to Ostrom's framework for analyzing common resources and the assumptions and principles that are the foundation of that framework. We also look briefly at the application of her analytical framework to urban *green* resources in cities as an indication of how well her framework might travel or translate in an environment vastly different from the communities she studied. Our bottom-line assessment is that Ostrom's framework offers a great deal to the study of common resources in cities. Indeed, we understand why it is tempting to apply Ostrom's design principles for user-managed resources to the management of many kinds of public and shared resources in the city. For many reasons, however, we think that Ostrom's principles do not work in the city exactly the way that they do in nature. Ostrom's framework needs to be adapted to the reality of urban environments that are often crowded, congested, socially diverse, economically complex, and heavily regulated. As such, we highlight some characteristics of *constructed* urban commons that are not captured well in Ostrom's design principles for user-governed natural resource commons.

Ostrom believed that there is no one model of resource management that is applicable to all common pool or shared resources. In some cases, collective or community governance is not the right solution at all and should give way to more traditional forms of state control or private property regimes. In her early Nobel Prize–winning work, she found that collective governance of shared resources is possible and even sustainable under certain conditions or when specific factors are present—what she called "design principles" (Ostrom 1990). Collective governance of shared resources is particularly successful for resources with clearly defined boundaries and in communities where it was clear who was "in" and who was "out"—that is, who had access and who could be excluded (principle 1). Ostrom's study of successful common pool resource institutions focused mainly on close-knit communities that share similar beliefs and a history or expectation of continued interaction and reciprocity. In actual field settings where these conditions are present, Ostrom observed that communities were able to develop and enforce rules as well as conflict resolution procedures that govern the use of the resource. She found that these communities had put rules in use that were well matched or adapted to local needs and conditions (principle 2). For this reason, she observed that rules of cooperation among users were written or modified by those entrusted with both the duty to obey them and the responsibility to enforce them (principle 3).

Collective structures and rules were premised on the assumption that communities' rights to self-govern the resource and to devise their own rules would be recognized and respected by outside central authorities; such recognition also made the rules easier to monitor and enforce (principle 4). For these communities, social control/monitoring and social sanctioning were two central pillars of Ostrom's design principles for the governance structure that communities often put in place to manage a common pool resource (principles 5 and 6). She observed that conflicts might arise because even the most united communities have internal fractures, and therefore communities require accessible, low-cost tools to solve their own disputes (principle 7). Ostrom found, however, that for more complex resources, its users were able to enforce and monitor the rules that they created only with the help of external agencies. Thus, the governance responsibility or decision-making power over the resources was shared with other actors to form so-called nested enterprises (principle 8). That is, the rules, procedures, monitoring, and sanctions put in place along with other governance activities are organized in a nested institutional structure with layers of activity by different actors. This nesting might occur between user groups using the same resource and/or between user groups and central authorities (e.g., local or regional governments). The involvement of central authorities is more likely for large-scale and complex resources. These are the basic design principles that for years have been driving the multidisciplinary study and observation of common, shared resources, namely, scarce, congestible, renewable natural resources such as rivers, lakes, fisheries, and forests (Poteete et al. 2010).

Many of Ostrom's cases studies documented the existence of wholly internal solutions to natural resources management. Ostrom identified a number of these self-organized resource governance regimes, including common lands governed by local village communities in Switzerland and

Japan (Ostrom 1990, 61–69), irrigation communities in Spain and the Philippines (Ostrom 1990, 69–88), and other examples of fisheries and irrigation projects managed communally in Turkey, California, and Sri Lanka (Ostrom 1990, 144–178). Many of these have survived for multiple generations and involved the investment of significant resources by participants to design basic rules, create organizations to manage the resources, monitor the actions of each other, and enforce internal norms to reduce the probability of free riding. Importantly, these groups successfully established and enforced their own rules without resort to external public agencies (Ostrom 1990, 59).

Ostrom's findings are consistent with similar research by others, such as legal scholar Robert Ellickson, highlighting the ability of small or "close-knit" communities to solve disputes over land use through a system of informal social norms (Ellickson 1991). Ellickson's study of ranchers and landowners in Shasta County, California, found that in spite of a well-developed system of legal rules that governed straying cattle and land disputes, the community had developed its own system of informal norms governing disputes and that the system was self-reinforcing. Ellickson's findings further support the idea that, at least in small homogeneous communities, the existence of strong cooperative norms allows communities to govern themselves in the face of conflict without the aid of the state or other central coordinator. How much the size of a community of users and its homogeneity affect the ability to organize and to self-manage a resource system is uncertain and requires more theoretical and empirical observation (Poteete and Ostrom 2004b).

For more complex and larger resources, however, Ostrom found that central regulators played a key role in helping to coordinate the interdependencies of smaller units of community-based governance (Ostrom 1990). Ostrom's study of a series of groundwater basins located beneath the Los Angeles metropolitan area is illustrative (Ostrom 1990, 103–142). In her findings, groundwater producers organized voluntary associations, negotiated settlements of water rights, and created special water districts to monitor and enforce those rights with the assistance of county and state authorities. State legislation authorizing the creation of special water districts by local citizens was a crucial element in encouraging users of groundwater basins to invest in self-organization and the supply of a local institution. Once a special district was created, it possessed a wide variety of powers. Those powers included the ability to raise revenue through a water pump tax and, to a limited extent through a property tax, to undertake collective actions to replenish a groundwater basin. Without such legislation, a similar set of users facing similar collective action challenges might not be able to supply themselves with transformed "micro institutions" (Ostrom 1990, 135). Ostrom viewed the relationship between the private water associations, public agencies, and special districts as illustrating how a governance system "can evolve to remain largely in the public sector without being a central regulator" (Ostrom 1990, 135). The basins became managed as a *polycentric* public-enterprise system that is neither centrally owned nor centrally regulated. As such, and in contrast to self-managed community resource use systems that operate mainly with social sanctions, resources that traverse many communities and/or heterogeneous user groups may require more complex institutional structures, often involving government coordination and enforcement (Ostrom 1990, 1994).

OSTROM IN THE CITY

For our study of constructed urban commons, many of Ostrom's design principles and observations are clearly applicable, and others are of limited utility or need to be modified to the urban context. For instance, we observe that, as in Ostrom's examples, the ability of communities to collectively manage a shared resource and to do so sustainably over time can very much depend on community size and cohesion, shared social norms/social capital, community homogeneity, resource scale, and recognition and support of central authorities and external actors. Similarly, collaboratively managed urban resources typically have clear boundaries and rules that are collectively created, adopted, and enforced through either informal or formal mechanisms. Unlike many of Ostrom's case studies, however, collective governance of urban resources does not occur only (or mostly) in small, homogeneous communities with stable membership and high levels of social cohesion. Many small and large-scale urban resources ranging from large urban parks to community gardens to wireless broadband networks are being collectively managed by heterogeneous groups of users who access and depend on the shared resource.

In addition to high group heterogeneity, many urban commons are accessible and open to transient users who are not part of a stable group of resource users who may be more geographically tied to the resource by virtue of their proximity to it. As Amanda Huron has noted, urban commons emerge in "saturated" spaces and often are constituted by the coming together of strangers (Huron 2015). Relatively high densities of population in a relatively small amount of space means that people are forced to either share or compete for resources, as Huron argues, making the process of urban "commoning" more challenging than in rural and small-scale environments. This is even more so in huge urban agglomerations that comprise growing core cities and expanding peripheries, including both formal and informal settlements.

As such, the role of central authorities or the state is even more present in the creation and sustainability of the urban commons and for reasons that differ in the natural resources context. As Ostrom argued, the effort by user groups to create new institutions for resource governance is a second-order collective action dilemma. In addition to overcoming any obstacles to cooperation to create rules of access and use, resource users must invest tremendous resources to design institutional arrangements that incorporate the new processes and rules that will govern the resource over the long run (Ostrom 1990, 136). This is why a small homogeneous community is more likely to succeed at managing a commons than a larger and more diffuse one. Apart from Ostrom's study of collective management of groundwater basins and the special water districts created to manage them, far less attention has been paid to the role of the state in the creation and support of user-governed or collectively governed resource regimes. As we noted in chapter 1, Ostrom's own work, as well as the work of others, suggests that central governments can play a significant role in supporting and potentially lowering the costs of user-managed resources.

The supportive or enabling role of government in the collective management of shared resources is unavoidable on some level in the urban context. Many urban resources that residents or communities want to share and manage together are, at least formally, under the control of the state. In many cases, the local government typically retains regulatory control and, in some cases, proprietary ownership of these resources. Communities and other private actors are motivated to claim, use, or preserve abandoned or underutilized urban resources as assets that can provide urban residents with essential resources such as affordable housing or commercial space, open and green space, and other goods. However, given that most of these resources are under government control and regulatory authority, users eventually need government consent and often government aid and financing to fully utilize the property. Thus, even for community-driven, constructed urban commons the state role can be essential to the creation and sustenance of these user-managed resources.

Because of the way that urban resources are controlled and regulated by central authorities, creating urban commons also depends on a level of legal and property adaptation above and beyond what is required to collectively manage or govern natural resources. In Ostrom's case studies of collectively governed natural resources, communities were managing true "commons" or common property alongside some private property rights to access those resources. Those common property rights in many cases were centuries old and coexisted with the development of private rights to those resources over time (Ostrom 1990, 63). Ostrom made clear that the common property being managed by communities is not the same as the "commons" open to everyone (res nullius) as conceived by many scholars since publication of Hardin's "Tragedy of the Commons." Rather, the common property regimes she observer existed "where the members of a clearly demarcated group have a legal right to exclude non-members of the group from using a resource" (Ostrom 2000, 335–336). These communities utilized natural resources and created the rules or conditions of access for themselves and others with built-in incentives for responsible use and sanctions for overuse.

Most urban commons are constructed from urban infrastructure as opposed to pre-existing resources from which users subtract (e.g., water or fish or wood). Cities are highly proprietary environments in which land and resources are often enclosed by ownership and exclusion rights that tolerate empty, abandoned, and unproductive "surplus" property to sit unutilized or underutilized for long periods of time. Creating urban commons most often requires changing or tweaking the way that public or private property is held and shared. In some instances, they require changing local laws to recognize or allow urban land and infrastructure to be used in common or creating new institutions that disaggregate and redistribute property rights and entitlements. As such, collectively governed, shared resources emerge as sites where self-organization takes place through "experimenting with rules by which to govern particular pieces of land and tinkering with the possibilities made available by existing laws and the features of private property" (Ela 2016).

In addition to the role of central authorities and property or legal adaptation, many kinds of urban commons are a product of what we call social and economic *pooling*. Scholars of the commons most often use the term "common pool resource" to denote the characteristics of an open access, depletable resources (Ostrom 1990). In this conception of a commons, the pool is the sum of the units that constitute the resource—for example, fish in a fishery or trees in a forest—and typically those units are limited and exhaustible. In economic terms, a common pool resource can be shared by many users simultaneously, but the amount or availability of the resource diminishes by every unit that an individual user subtracts. The pool is thus depleted or exhausted when too many unconstrained users have taken from it, leading to the classic *tragedy of the commons* scenario.

Our use of the term *pooling* is not to denote existing, open-access, non-renewable units. Rather, pooling is the combined effort and associated resources of different actors to construct and share common goods. Community gardens, wireless networks, co-housing, and land trust arrangements are most often the result of pooling human capital, social networks, and existing urban infrastructure or public resources to create and or construct shared resources. These resources are then made available and accessible to a broader class of urban inhabitants, many of whom are on the social and economic margins of growing cities. Resources become an urban *commons* or part of a common *pool* through these collaborative practices and ventures aimed at sharing existing urban resources, generating new resources, producing new public services, and coordinating urban networks across the city.

GOVERNING GREEN URBAN COMMONS

One entry point in assessing whether and how Ostrom's approach applies in the urban context is to ask whether natural ecological resources are being collectively governed or managed in cities under conditions like those found in Ostrom's case studies of traditional common pool resources. These urban ecological resources or "urban green commons," which include lakes, parks, and urban gardens, can provide critical resources such as food and recreational spaces for urban populations that live near them (Colding et al. 2013). They can also be important spaces that strengthen social networks and facilitate social integration in dense, diverse, and often socially stratified urban environments. These resources can be as vulnerable and endangered as natural resources in rural environments and perhaps even more so because of urbanization patterns. For instance, urban green commons are frequently privatized, converted to built spaces, degraded, or polluted (Unnikrishnan and Nagendra 2015; Mundoli et al. 2015). Like traditional common pool resources, they are also subject to rivalry and conflicts with respect to their use, management, and ownership in urban environments characterized by rapid urbanization, migration, and landscape change (Unnikrishnan et al. 2016).

Collective action to manage these resources in cities can mimic, at least at first glance, similar resources in the natural world or in rural areas. To test this, researchers have applied Ostrom's framework for the study of the commons in cities by examining ecological resources such as lakes, rivers, and forests accessed by urban local communities for traditional cultural and livelihood uses and/or by recent urban migrants for aesthetic and recreational purposes (D'Souza and Nagendra 2011). Ostrom developed an institutional analysis and development (IAD) framework to analyze collective action situations with a focus on institutions in which multiple actors are interacting (Ostrom 2005a, 2011). The IAD framework includes both *endogenous* (internal) and *exogenous* (external) variables that can influence how well a particular resource is being collectively managed by a local community (Ostrom 2005a). These include the biophysical characteristics of the resource, attributes of the community, rules in use, the action area or arena where participants interact and solve problems (or not), and information about specific actions situations and specific actors (Ostrom 2005a). The IAD framework was later expanded into Ostrom's social-ecological system (SES) framework involving a set of ten variables that include the size of the resource system, number of actors, leadership, social capital, importance of the resource, existence of operational-choice rules, and existence of informal mechanisms for monitoring (Ostrom 2007, 2009a). The IAD and SES frameworks have been used by scholars to examine case studies of lobster fisheries, forests, irrigation systems, grazing pastures, and other scarce, congestible, nonrenewable natural resources (Poteete and Ostrom 2004a). It has also been used and adapted to study collective governance arrangements for other kind of resources such as knowledge and cultural commons (Frischmann et al. 2014).

In the first robust application of Ostrom's approach to natural resources in urban areas, Harini Nagendra and Elinor Ostrom examined the challenges of collective governance of urban lakes on the periphery of Bangalore, India, using the SES framework's social-ecological variables associated with self-organization in previous studies of traditional commons (Nagendra and Ostrom 2014). These variables were applied to lakes of varying size and ecological quality (from lightly to very polluted) located on Bangalore's urbanizing peripheral areas to diagnose why some water bodies had been effectively restored and managed by newly forged collaborations between citizens and local government locations, whereas others had become ecologically deteriorated and/or failed to generate sufficient levels of collective action. Consistent with Ostrom's observations of traditional common pool resources, the study of urban lake commons found that endogenous factors were very important to the presence of collective management. Specifically, collective action was present in six of the seven lakes studied, where the following variables were present: a small or moderate number of actors, the presence of local leadership, relatively high levels of trust and social capital, lack of exclusion of socioeconomic groups, high resource importance to residents, and the presence of operational community rules and informal norms for monitoring the resources. Yet, those collective efforts alone were unlikely to have improved the ecological condition of the lakes, some of which were very polluted. Rather, the study found that it was the *combination* of endogenous and exogenous factors that correlated with a high level of collective action *and* high ecological performance. Notably, only two of the six lakes were characterized by *both* collective action and improved ecological conditions.

Most important for our purposes is the authors' conclusion that the challenge of cleaning up an urban lake in a quickly urbanizing area on the periphery of Bangalore required effective interaction or collaboration with various governmental units and other actors (Nagendra and Ostrom 2014). Collaboration and networking with others are critical in the urban context, the study stressed, because of the complex legal, technical, and political environment in which these lakes are located. For instance, lake restoration requires technical, financial, and manpower resources necessary for the tasks of dredging, bund building, and other cleanup activities that are beyond the scope of local resident groups to manage alone. Local resident groups must work with government agencies as well as technical experts (e.g., researchers and naturalists) to successfully restore the resource to a level that can meet their local needs. At the same time, although government agencies have the legal authority to prevent unwanted activities and harmful use of the lake, they must rely on information from local residents to detect these activities and intervene in a timely manner. Collective action by local groups is not only critical in monitoring the process of restoration and ensuring that the lakes remain in healthy conditions after rejuvenation. Such collective action is critical also, the authors conclude, in strengthening downward accountability (ensuring the effectiveness of monitoring against infractions and sanctioning of repeat offenders) because local officials are not always accountable to the residents they serve given the economics of urbanization and the imperatives for growth in many cities. As the authors note in this context, "local officials are often subject to governance incentives as well as incentives of political economy and rent-seeking that ensure that they are primarily accountable to higher officials or vested interests such as real estate agencies, rather than downward accountability to local communities or marginalized groups" (Nagendra and Ostrom 2014, 76).

ENABLING URBAN COMMONS

Much like ecological commons such as lakes and rivers, constructed green commons such as neighborhood parks, community gardens, or urban farms must account for the political, economic, and legal complexity of the urban environment in which they are located. Endogenous efforts alone are rarely enough to maintain or sustain over the long run collective efforts to manage or govern even small resources such as community gardens and urban farms. These collective efforts of local users ultimately depend on some cooperation with central authorities—that is, local government officials, administrative agencies, and others responsible for managing and governing different kinds of urban infrastructure. At the same time, local collective efforts to manage these resources are vulnerable to the larger urban political economy in which these efforts are situated. For these reasons, the economic and political complexity of cities, including rising social and economic inequality, means that governance of urban commons is often not just about communities governing themselves. Rather, the creation of new urban commons almost always involves some form of enabling or support from the local government or state and, in most cases, cooperation with other urban actors and sectors. However, what degree of state enabling is necessary for sustainable collective governance of shared urban resources and how vulnerable these resources are to capture by a narrow set of interests that are not fully accountable to the surrounding community or to co-optation by extractive market forces or private actors, are heavily dependent on local context.

Consider the example of community gardens, one of the most ubiquitous kinds of urban commons in cities around the world. The transformation of vacant or abandoned land into productive urban resources is initially an endogenous effort in which residents self-organize and self-manage these spaces as shared community resources (Foster 2011). Residents manage to come together, clean up or restore the lots, and construct and maintain fully functioning urban gardens and farms. Local users collectively formulate their own rules of use and allocate resource units (e.g., plots of land) and shared infrastructure (e.g., water connection or greenhouse) without a formal organizational structure (Rogge and Theesfeld 2018). Constructing and maintaining community gardens and farms often depends upon and fosters collaborative relationships and social ties among residents of different neighborhoods and racial and generational groups (Foster 2006). This social capital and the norms that they generate enable residents to cooperatively work toward common neighborhood goals and a shared desire that the space serve the needs of local residents—whether providing fresh vegetables, green space, or recreational amenities. Moreover, there is evidence that these self-organized efforts tend to spread throughout urban areas through a "social influence" or "social contagion" process (Shur-Ofry and Malcai 2019). In other words, the creation of community gardens at the micro level or sublocal scale enables and supports the diffusion of these efforts on a larger scale on a citywide basis. This diffusion and contagion occur through the interactions among individual participants or players from different community gardens facilitated by enabling nudges—positive reinforcement or supportive programs—from central authorities.

The role of central authorities or regulators can be important in both enabling and sustaining locally organized efforts both by providing modest support and assistance to these users (Lehavi 2008; Foster 2011). Abandoned, vacant and underutilized spaces on which community gardens or urban farms are constructed, for example, are most often under the control of central authorities. They can operate long term as community gardens or urban farms only with the implicit or explicit consent of the local government. Sometimes city officials may passively allow the group to utilize land under the city's control and refrain from interfering in the group effort. Other times, city officials might transfer land to the group either for a nominal fee or for a contractual term and may even provide materials and other critical resources to the gardeners, such as access to gardening equipment through city gardening programs. New York City's GreenThumb Program is an example of this kind of support, providing residents with technical support and materials (Foster 2011). Local land use rules and zoning might have to be changed to allow for a change in the use of land from a former residential or commercial use to its current agricultural use. Residents might also need to take advantage of local rules and regulations on access to local water supply and other urban services or infrastructure required to engage in urban gardening or farming (Ela 2016).

This enabling role grows more significant with the scale and complexity of the resource, involving the need for much more legal authority and/or financial entanglement than smaller resources require for collective management. A stronger state role is required when resources are not only larger but involve more heterogeneous users and involve more legal or regulatory complexity. For instance, as with community gardens, collective efforts to revitalize and manage neighborhood urban parks are largely endogenous efforts undertaken by abutting park neighbors or frequent users who lend their time, give money, or help raise funds to recover and maintain the park. These groups consist of volunteers, typically referred to as Friends of Park [X], who provide labor for park maintenance and assist in community outreach and park programming. They organize park cleanups and community events, build or donate simple infrastructure or facilities for community activities (e.g., small pools or sand pits), and patrol the park as a way of deterring criminal and other undesirable activities (Madden et al. 2000). Many of these groups remain an informal collection of volunteers, whereas others have become more formal. The more formal groups establish themselves as a membership organization, elect board of directors, write bylaws, and apply for nonprofit status (Lehavi 2004).

These community-based Friends of Park [X] groups tend to rely heavily on government assistance, and in some instances collective efforts are very much dependent on the government to coordinate, establish, and sustain these efforts. Local governments help to develop and nurture these groups by providing them with technical assistance, training, and funding (Madden et al. 2000). An example is New York City's Partnerships for Parks, a joint venture between the New York City Parks Foundation and the New York City Department of Parks & Recreation, which encourages the formation and nurtures the development of neighborhood parks groups across the city. The provision of training, materials, and financial support to local groups willing to assume some responsibility for some park management functions can provide a powerful signal and incentive for individuals to pool and coordinate their efforts as well as sustain the enterprise over time. Like community gardens, this state enabling role is crucial even when there are strong endogenous factors at play that enable communities to engage in collective action to care for a shared urban green resource. These efforts may not be successful nor sustainable over the long run, despite strong social ties and cooperative action, were they not assisted by local governments through local programs like those mentioned.

In our previous work we have observed that state enabling of self-organized, collective governance of shared urban resources exists along a spectrum (Foster 2011). Enabling mechanisms range from offering de minimis support to largely endogenous collective efforts, as in the examples of community gardens and small neighborhood parks, to more significant support in which central authorities are essential to the formation of collective efforts. At the de minimis end of the spectrum, central authorities allow, either explicitly or implicitly, the collective to exercise management prerogatives over the resource and may offer them material support to start and sustain their efforts. The government has virtually no affirmative role in coordinating the collective effort or in establishing the group, although it may provide them with financial or other incentives to sustain their efforts. Further along the spectrum, there can be a closer relationship between central authorities and the collectivity in which the government shares its resources with the group and exercises some degree of oversight of the group's activities. Government enabling is an important stabilizing force for the group, and the group works closely with government officials. However, the relationship between the government and the group falls short of a fully realized partnership. On the far end of the spectrum are collective efforts that are very much dependent on the government to coordinate, establish, and sustain them. That is, the group takes its form only as a result of government support and entanglement, and government support is a precondition to the existence of collective action.

Two examples of the latter kinds of larger-scale state-enabled, sublocal governance arrangements are park conservancies and business (or community or neighborhood) improvement districts. *Park conservancies* are constituted of public and private stakeholders who maintain and manage, in partnership with city government, large urban parks. In contrast to park "friends" groups formed to support small neighborhood parks, park conservancies are nonprofit entities that raise significant amounts of money and co-manage large urban parks in partnership with the local government by collaborating on planning, design, and implementation of capital projects as well as sharing responsibility for park maintenance and operations and in some cases revenue (Taylor 2009; Murray 2010). The prototype for park conservancies is the Central Park Conservancy in New York City, which was founded by several local leaders and groups that initially established the Central Park Task Force, an organization that began to encourage direct involvement of the public as park volunteers and donors but later incorporated itself as the Conservancy. In a groundbreaking power-sharing arrangement, the Central Park administrator was appointed to serve as the chief executive officer of both the park and the Conservancy, signaling the important role that the Conservancy would have in the restoration and maintenance of the park. The Conservancy is run by a board of trustees, which includes city officials and representatives from nonprofit organizations and private corporations, among other interests. It combines donations from individuals with corporate donations and government funding to fulfill its budgetary needs and build its endowments. A variety of public bodies have oversight over the Conservancy's management decisions, including the Art Commission of the City of New York, five neighborhood community planning boards in the city, the Landmarks Preservation Commission, and the city council. Although Central Park Conservancy may be the most widely known of park co-managers, its model has been widely replicated with varying success in large urban parks around the US (Taylor 2009, 350; Rosenzweig and Blackmar 1992, 524).

Agreements or partnerships between local governments and park conservancies serve an important coordinating and stabilizing function that enables disparate sectors and groups to cooperate to undertake significant responsibility for park management. Private involvement in the management of urban parks is a phenomenon stretching back to the early twentieth century (Kinkead 1990). Neighbors that live near urban parks, as well as wealthy donors and residents, have long exerted some power over park management by providing donations, labor, advocacy efforts, and planning ideas. Often, however, these efforts have suffered from a lack of coordination and efficiencies of scale; without leadership to harness these private efforts, they often falter over time as old groups fade and new ones appear to renew the effort to resuscitate and improve park management (Murray 2010). Agreements such as the one between the city of New York and the Central Park Conservancy serve both to establish important norms regarding the limits of the group's responsibility for the resource—that is, reverse crowd-out protection that ensures public funds will not be replaced by private donations—and to formalize the contours of the conservancy's responsibility for the day-to-day management of the park. These park conservancies have been widely credited for the revitalization of urban parks at a time when some cities had "all but abdicated their role as stewards of the public parks" (Taylor 2009, 346–347). They have the virtue of being able to avoid the red tape, bureaucracy, and inaction in which city parks departments often become mired; they can make decisions faster, raise funds, save money, and serve as effective advocates for urban parks.

Similarly, partnerships between local businesses, property owners, and local governments are established to manage the neighborhood commons—that is, streets, sidewalks, parks, and playgrounds. Business improvement districts (BIDs) are enabled by state and local legislation that allows a majority of commercial property owners in a defined neighborhood to vote to form a BID, agree to pay special assessments, and assume (at least partial) control and management (maintenance) of the neighborhood commons. BIDs are governed by local property owners in partnership with representatives from businesses, local governments, and sometimes neighborhood resident non–property owners. The key features of BIDs are that (1) they cover a defined (and limited) geographic territory in which commercial property owners or businesses in the area are subject to additional assessments or taxes; (2) they typically fund supplemental street-level services and small-scale maintenance and capital improvements (e.g., street cleaning, garbage collection, landscaping, sidewalk widening, and security patrols) over and above those offered by city government; and (3) they are granted the limited authority by legislation (Briffault 1999). Similar districts have been established as community improvement districts (CIDs) or neighborhood improvement districts (NIDs), mostly to encourage and fund economic development and public improvements in defined neighborhoods through collecting special assessments from property owners or imposing special sales or license taxes in the district. These special districts are now a ubiquitous feature of urban governance in many cities across the world, with varying governance and financial arrangements.

Because BIDs exist only by virtue of specific legislative authority, enabling legislation is what allows local commercial business and/or property owners to minimize free-rider and coordination problems in order to provide neighborhood services beneficial to the local environment. BID legislation (and similar legislation authorizing NIDs or CIDs) lowers collective action costs by arranging for the municipality to collect the mandatory assessment from property owners who then use the funds to provide services. The impetus for a BID creation typically arises from a significant portion of the property owners or businesses in the neighborhood, or representatives of one or more of those groups, that organize the BID and agree to assess themselves or impose a sales tax or other tax in order to fund the activities and services provided by a BID. BID formation is often costly in terms of time, energy, and money to coordinate and prepare the necessary groundwork, and it can take years before the process is complete (Briffault 1999). BID legislation can enhance the capacity to achieve collective outcomes among diverse actors, even in the private sector, whose interests may not appear at first to be well aligned.

These special institutional arrangements mimic to some extent Ostrom's findings on management of regional water basins through special water districts. Special water districts were legislatively enabled to make possible collaborative water governance involving groundwater producers, residents, and state and county authorities. As in her findings, the state or central authorities play a key role in helping to enable and coordinate these nested units of resource governance for larger and more complex resources. In the case of park conservancies, the local government helps to establish them and becomes part of a formal partnership to collectively manage the resource with private and (sometimes) community-based actors. In the case of BIDs, the state must enact special enabling legislation to establish them, including defining their authority and fiscal responsibility over common shared neighborhood resources. However, one difference between the natural and the urban environments is the political and economic context in which state enabling occurs. In urban environments, they occur in an often highly unequal context that includes race, ethnicity, and/or class segregation and stratification. As such, although the role of the state in enabling collectively managed large parks and neighborhood common spaces is largely seen as a positive, it has also come under criticism for the ways that these nested institutions exacerbate distributional inequalities in public goods and services.

Our view is that these kinds of self-professed *public-private partnerships* can carry costs for urban communities least able to participate in the stewardship of these common resources that they manage. Park conservancies, for example, have been criticized for imposing many of the costs that attend to the (at least partial) privatization of any public good—that is, enabling gentrification, exacerbating ethnic and class tensions, and creating a two-tiered park system that disadvantages parks in less affluent neighborhoods (Taylor 2009; Murray 2010). Enabling the partial privatization of large urban parks or entire neighborhood common areas might result in the creation of different tiers of common resource stewardship, depending on the demographics of those who live closest to the resource and/or frequent it the most. Although local government enabling is available to any group of private actors able to overcome free-rider and other collective action obstacles, the scope and success of the management or stewardship effort will depend in no small part on the assets of those individuals involved (as well as their ability to attract additional assets). Although park conservancies are celebrated for raising and dedicating private funds toward the improvement of larger prominent city parks, they often create a two-tiered park system that disadvantages parks in less affluent neighborhoods. One cost of their success is that parks and playgrounds in poorer neighborhoods are often left underfunded and relatively unattended (Taylor 2009, 302).

In a similar vein, BIDs are widely credited with making small-scale improvements to streets, parks, and other common areas that have led to the revitalization of once deteriorated urban commercial areas like New York City's Times Square. However, BIDs raise concerns about the extent to which they exacerbate the uneven distribution of public services. BIDs in low-income neighborhoods tend to have less fiscal and human capital (because of lower property values) to dedicate to street-level services and capital improvements than do those in high-income neighborhoods (Gross 2005, 184). Less central or popular parts of the city, without the support of wealthy private partners or commercial businesses paying premium tax rates, suffer from underfunding because of the success of other, more visible areas of the city. The result is that the BIDs in these neighborhoods provide a very limited range of services that tend to address the most visible aspects of urban decay (e.g., graffiti, sanitation, and sidewalk maintenance) and fall far short of the kind of major capital improvements that characterize BIDs in central downtown or wealthier neighborhoods. The governance structure of BIDs has also been challenged, both in academic commentary and in the courts, for lacking democratic accountability and in part for its exclusion of non-property-owner residents from participating in BID management of their neighborhood (Foster 2011). Moreover, once they are established, there is in fact very little oversight of them, even though most BID legislation provides the authority for oversight by politically accountable government officials (Briffault 1999).

BOTTOM-UP VERSUS TOP-DOWN URBAN COMMONS

Park conservancies and BIDs are one form of urban commons, or collectively governed shared urban resource. They represent, however, a top-down, institutionalized, state-enabled form of collective resource governance. They are top down because they are initiated and come into being only through government authority or action, as is the case of large park conservancies and BIDs. On the other hand, many other kinds of urban commons emerge from bottom-up efforts of residents or resource users who are motivated to overcome traditional collective action problems and to collaborate to construct new goods and services that many urban communities lack or find inaccessible to them. The issue for bottom-up urban commons is not only determining what is the best way to manage or govern existing resources like parks, land, or existing urban infrastructure. Rather, the greater issue is how new forms of urban commons can emerge from those resources that are already under some form of legal ownership and control, whether public or private. The challenge is how communities can access and utilize existing resources and urban infrastructure to *construct* new resources and goods that respond to community needs but that are under neither exclusive public nor private control.

Cities are highly proprietary environments, as we have previously noted. Land and structures that are not privately owned are public property of some sort, meaning that they are under the control of the state (local government or higher levels of authority). Public property can include streets, roads, squares, parks, cultural institutions, and other structures dedicated to public use. However, one question that arises in cities all over the world is whether private or public property that is abandoned or vacant should be potential sites for urban commons. As we mentioned in the previous chapter, cities and neighborhoods characterized by growth and those characterized by shrinkage and decline contain significant amounts of vacant land and empty or underutilized structures. Land or structures in cities can become vacant or underutilized for many reasons, depending on whether the resources are public or private property. In some cases, public buildings owned by the state may fall into disrepair or disuse due to lack of public moneys to take care of them. In addition to underutilized or vacant public land and structures, private land and structures can end up in the public domain when owners default on their tax obligations or otherwise abandon the obligations of property ownership. Local governments in many cities assume responsibility over these parcels, sometimes actively through tax foreclosure and sometimes by default. They become, at least temporarily, a form of public property while in the public domain.

In this transitory state of moving away from a past use and toward a future use that is unknown and unplanned, vacant land and structures are quite vulnerable to contestation of uses. Conflicts often emerge regarding present vs. future uses and different possibilities for future use. These conflicts exist between present owners of the land and the local government, and between the surrounding community and the local government, which may be hoping to sell abandoned property to private developers or investors. There are also conflicts among various users who have or gain access to the property and who may have in mind competing uses for the property. In some communities, residents are treating vacant land or abandoned structures as an open access resource to be shared broadly and utilized to produce goods for the community. As such, community members may begin to treat the property as an open access resource, utilizing it in ways that add value to the surrounding community and/or which produce goods for that community (as in the case of community gardens or urban farms or using abandoned homes to house the homeless). In other instances, public users conduct illegal activities such as dumping or crime, which clearly does not add value to the surrounding community. In fact, as we have previously argued, the rivalry in these spaces could lead to an urbanized version of "tragedy" in which open access leads inevitably to further degradation or destruction of the shared resource.

Pushing against this tragedy narrative for vacant and abandoned spaces is another narrative rooted in the language of the commons. Unlike

Hardin's tale of tragedy in these spaces, opening up access to abandoned or vacant property instead can enhance and capture positive value for the community by virtue of using the property to create goods (both tangible and intangible) that can be shared. This narrative or argument characterizes several social movements in the US and abroad in which activists occupy vacant, abandoned, or underutilized land, buildings, and structures. These movements are responding to what they view as market failures and the failures of an urban development approach that has neglected the provision of goods necessary to human well-being and flourishing. The tactic of occupation is a form of resistance against the enclosure, through private sale or public appropriation, of these assets or property in transition. Occupation is also a way of asserting that the occupied property has greater value or utility as a good either accessible to the public or preserved and maintained as a common good.

For example, in many parts of the US as well as in countries such as Brazil and South Africa, individuals occupy and squat in foreclosed, empty, often boarded-up homes and housing units (including public housing units) as a means to convince municipalities to clear title and transfer these homes and units to limited equity forms of ownership in order to provide long-term affordable housing for neighborhood residents (Alexander 2015). This "occupy" or "take back the land" movement is a response to the displacement of homeowners and tenants brought on by the confluence of the housing/mortgage crisis and the forces of gentrification. Rather than leaving these homes vacant and blighted, local public officials often condone the occupation and transformation of these structures by community members who aim to return the asset to productive use in ways that beautify and improve the properties and by extension the surrounding neighborhood (Alexander 2015, 271).

In a similar way, the Italian movement for *beni comuni* (common goods) has used occupation to stake public claim to abandoned and underutilized cultural (and other) structures in an effort to have these spaces either retained as or brought back into public or common use (Bailey and Mattei 2013). The most famous of these occupations occurred when a collection of art workers, students, and patrons occupied the national Valle Theatre in Rome (Bailey and Marcucci 2013). The theatre had become largely defunct as a result of government cuts for all public institutions, and the Italian cultural ministry transferred the management of the theater to the city of Rome. Out of fear by many that the city would then sell it to a developer as part of a larger project for a new commercial center, a collection of art workers, students, and patrons occupied the theater. This occupation was followed by similar occupations of theaters and cultural institutions that were subject to privatization in cities all over Italy. In each case, the occupants' aim was "to recover people's possession of underutilized" structures and "open up" these spaces for the flourishing of common goods like culture (Bailey and Marcucci 2013, 997).

Although not explicitly using the language of the "commons," these contemporary "property outlaws" (Peñalver and Katyal 2010) were very much staking claim to vacant, abandoned, and underutilized land and to structures as common goods that should be accessible to urban dwellers to create essential resources for their communities. As Peñalver and Katyal's work has demonstrated, those excluded from property often respond in ways that end up reshaping legal norms on property ownership and use. From "illegal" lunch counter sit-ins during the civil rights movement to selective online copyright infringement, "property outlaws" often strengthen the role that property should and can play in changing the legal and social order. Although the creation of urban commons does not turn per se on outlaw activity, the claiming of underutilized land, structures, and other urban assets challenges the public/private binary of property ownership in which either the state or private actors have sole and exclusive dominion over urban property. In other words, occupation becomes part of an effort to transform a strictly private or public good into a *common* good, made accessible for sharing and possession by a group of local inhabitants.

Consider the way that a collective group of artists in Milan has drawn public attention to the amount of unused and underused spaces in the city and helped to push the city council to recognize the value of utilizing abandoned private and public property to meet community needs (Delsante and Bertolino 2017). By squatting in abandoned property and remaking those spaces for everyday cultural and artistic activities, this collective has advanced its underlying goal to "promote a dialogue with institutions to recognize the process by which an abandoned space could be considered a common-pool resource and thus be made available to the community" and "directly managed by self-organised groups of what it calls 'active citizens' through processes of participatory democracy" (Delsante and Bertolino 2017, 53). On the heels of these occupations, and following substantial political debate, the city provided the collective access to some vacant properties and issued a larger call for proposals to temporarily use available spaces around the city. This was a precursor to the passage in 2012 of a city ordinance setting out criteria allowing the "re-use of vacant spaces" and unused buildings, both public and private, for "for the development of artistic, social and economic activities." A broad class of users, both public and private, could take advantage of this ordinance. The ordinance specifically identified on an "experimental basis" a list of spaces proposed by any citizen or group of citizens to be used in the public interest free of charge for a maximum of three years with the possibility of renewal (Delsanti and Bertolino 2017, 52). What happened in Milan is reminiscent of Ostrom's observation that successful user-managed or collective governance of a shared resource is recognition and respect by higher-level authorities.

As we discuss in the next chapter, several cities have adopted policies or regulations that acknowledge or even enable the use of publicly controlled or owned land or buildings for the creation of common goods and services by a collective group of citizens or users. This willingness is particularly evident in Italian cities but is spreading to other cities on the European continent through policy diffusion—learning from the experience of Italian cities and adapting those policies within their own legal, social, and economic contexts.

COMMONING IN THE CITY

Policies such as the one developed in Milan recognize that those involved in the creation of urban commons are not simply creating new kinds of resources but also new community-based institutions for sharing those resources. In this respect these policies recognize, at least implicitly, the value of what many scholars refer to as *commoning*. As prominent commons theorists David Bollier and Silke Helfrich argue, the commons is not only about resources, goods, and things but also about an ongoing social process and practice involving human interaction and social relations within communities—whether they be physical or digital communities (Bollier and Helfrich 2015). Bollier and Helfrich understand commons, as we do, as a blend or co-mingling of a physical (or digital or natural) resource with "social practice and diverse forms of institutionalization" (Bollier and Helfrich 2015, 6).

Creating new commons, or *commoning*, is about a set of practices and sometimes institutions that aim to decommodify resources and resist traditional norms through collaborative organization and decision making (Bunce 2015). Commoning thus describes the bottom-up practice of collectively creating or constructing resources that can be shared with others and that meet concrete user needs. It requires not only a resource around which to common but also a community that has access to and can take care of, manage, and steward the resource over the long run. Commoning begins with the internal work that a community of users must do to create new common goods and then expands to develop the capacity for collective management of existing resources on the basis of strong cooperative norms and shared goals. In this sense, commoning is highly pragmatic, involving the establishment of rules and conditions and in some cases institutions for collectively sharing resources among a defined social group or group of users. The practices and patterns of commoning vary, of course, depending on the resource, the nature of the community of users, and the social or cultural context.

We caution that it is easy to romanticize the idea of commoning and of communities coming together to form relationships or build on existing relationships and to collaboratively create and then govern resources they require to flourish and improve their communities. This is time-consuming and hard work, sometimes exceeding the capacity and/or the desire of communities and users to undertake it. As Ostrom's research demonstrates, commons are not solutions for all social problems, nor can they exist or function sustainably under all circumstances. As her design principles reflect, the hard work of commoning may be possible only under certain conditions such as small homogeneous communities and resources with clear boundaries. Even with state enabling, as we have described, local actors need incentives and sometimes significant external support to engage in collective governance, constructing and creating new resources out of existing ones, and then managing them sustainably over time. This can seem even more daunting, but not impossible, in urban environments that are large, dense, socially heterogeneous, and economically competitive.

Nate Ela's sociolegal mapping of community gardens in Chicago also reminds us that commoning in urban environments occurs in a particular legal context that shapes how commons institutions are constituted (Ela 2016). His study of how people in two neighborhoods on Chicago's South Side gained access to and sought to govern land for community gardens and urban farms reveals that self-organizing occurred in relation to the rules created by state and local government. In the case of gardens, Ela emphasized the ways that individuals and small groups were iteratively searching for ways to secure use and ownership rights over land and its products. Claiming access rights to a particular space or plot of land and governing its use requires more than strong social norms between strangers. It also can mean navigating a thick layer of laws and regulations that need to be realigned with recognition of the commons as a form of property stewardship in order to institutionalize collective governance of those resources.

Amanda Huron's rich account of tenant organizing in Washington, DC, to create limited-equity cooperatively owned housing is another example of some of the dynamics involved in urban commoning. Huron recounts the story of hundreds of residents across the city who found themselves faced with eviction notices to make way for the razing of their structures in order to build tall luxury condominium apartment buildings in a quickly changing city. The DC residents, mostly low- or moderate-income African Americans and other minorities, were vulnerable to eviction at a time when middle-class residents were returning to centrally located, historic city centers (Huron 2018). For years, tenants across the city worked together to fight their evictions, to pool their money to purchase their apartment buildings and remain in place, and to exercise control over the increasingly scarce resource of affordable housing. To ensure the affordability of these buildings for future low- and moderate-income persons, the current residents created limited-equity cooperative ownership structures. This structure allows apartment dwellers to purchase shares in the co-op for little money, to pay low monthly co-op fees, and then to sell their shares for the same amount that they paid for it plus a small amount of interest. To create and sustain this collectively owned and controlled resource, however, residents who were often strangers to each other (and did not even speak the same language) had to create their own governing structures, negotiate with city officials, find financing, work together to repair and remodel their buildings, write bylaws for making decisions, and decide on house rules and rules of access and exclusion (i.e., who is and is not allowed to buy into the co-ops). Despite the seeming barriers of culture and language, these "strangers" were able to claim and create a common resource together—in some instances, even holding their meetings in as many as three languages (Huron 2018, 87).

Huron describes the creation of these urban commons as "unintentional" in the sense that the residents involved were not seeking to create common-interest communities nor to create a new institution to democratically govern themselves and their shared resource. These were essentially strangers coming together—tenants who happened to live in the same community—who were compelled to respond to a housing crisis under intense pressures of time and money. "It is about creating spaces not just for the people members know and love—though, as seen, this is certainly an important part of it—but for people they don't yet know, perfect strangers tossed their way by the currents of urban life" (Huron 2018, 160). They did not start out necessarily wanting to engage in commoning or even appreciating what would be involved. They came together for pragmatic reasons and sustained the practice out of collective need. Commoning is one option among a limited array of options, Huron argues, in cities that have become sites of intense capital accumulation. Traditional home ownership is not an option for this class of residents. As such, she argues, for people without access to capital, commoning is rational economic behavior. Although, clearly, LEC members are benefitting financially, their economic self-interest is not driving the creation of the commons. Rather, constructed commons like these create and support economies that are collaborative in the sense that community stability, control, and affordability are important elements as well.

The Bin-Zib co-housing communities in Seoul, South Korea, similarly illustrate the dynamics and work of creating urban commons in dense and heterogeneous environments as opposed to the small homogeneous environments that were the focus of Ostrom's studies. It also hints at the potential for urban commoning to scale across a city through networks of strangers motivated to work together toward a more collaborative and regenerative social and economic system in their city. Founded in 2008 by three people in their early thirties, Bin-Zib started as a communal living experiment in a three-bedroom apartment open to share with other "guests" for any length of stay (Han and Imamasa 2015). After purchasing the apartment, the founders invited others to live there; they disavowed any "ownership" in the house but accepted contributions by those who passed through according to their ability to pay (although everyone paid an equal amount for shared living expenses). There were no rules for membership in Bin-Zib, allowing people to come and leave as they wanted for any length of stay. As the number of guests increased, so did the number of rented houses that became part the Bin-Zib network of houses, which grew to over twenty houses over the years (although many of these have been disbanded) (Han 2019). The Bin-Zib has grown not only in size but also in its impact. The inhabitants open new houses when the existing ones become congested, and each house is managed or governed according to the social norms and relations of its occupants.

Bin-Zib is a unique and potentially replicable example of commoning in a heterogeneous, congested urban environment. Because these houses are open to anyone, they attract a cross-section of people with different motivations, ideologies, and sensitivities. There is documented conflict but also "convivial socialization," including frequent online discussions, parties, and collective events that promote relations between community members and allow for the constant re-articulation of what Bin-Zib means to the community (Han and Imamasa 2015). Notably, Bin-Zib has no articles of association, maintains a flexible structure in which everything is decided by discussion, and intentionally keeps the community as open and heterogeneous as possible to further the aim of preserving its egalitarianism (Han 2019, 181–182). Ben-Zib's development has been a process of trial and error in which "[t]he community has changed through solving specific problems residents have encountered, and the ways of solving problems were, in many cases, spontaneous" (Han 2019, 177). Bin-Zib has grown into a network of houses around Seoul, and it also includes a community café and a community bank that supports Bin-Zib communities and several other co-housing communities around the country (Han and Imamasa 2015; Han 2019). In some of the communities where Bin-Zib exists, its residents actively network with other local actors such as individual artists, religious groups, and merchants' committees to support the emergence of a more collaborative and regenerative local economy (Han 2019, 186–187).

The Bin-Zib example highlights another unique aspect of urban commoning, what we refer to as social and economic *pooling*. Pooling recognizes the capacity of multiple urban and local actors—for example, city inhabitants, civil society organizations, local businesses, social innovators, and knowledge institutions—to access and use existing urban infrastructure to generate new resources (goods and services) that meet community needs (Iaione and De Nictolis 2017). By mixing and matching social and economic resources dispersed across the city, pooling expands the capacity of these existing resources and of the participants involved in sharing them. Social and economic pooling blends individual and organizational capabilities and occurs across economic and institutional boundaries, often filling the spaces or voids in the access and delivery of essential goods and services. Further along in the book, we return to the ways that pooling can be supported and enabled by city policies.

LEGAL AND PROPERTY ADAPTATION

What ultimately makes the creation of new forms of shared and collectively managed urban goods challenging is the cost and access to urban land and infrastructure. Urban land and various kinds of urban infrastructure are increasingly a vehicle for high investment returns and the target of public and private efforts to capture and exploit their market value. Gaining and/or retaining access to these resources often involves a struggle or effort to recognize something akin to a collective property right to those resources for the urban poor (Blomley 2008). As the opening anecdote in chapter 1 on the vulnerability of community gardens reveals, rising land and real estate values threaten to displace longstanding communities from the material and immaterial resources, such as social networks, on which they depend. This threat looms even for newly constructed communities like Bin-Zib that are beginning to see rising rents in their neighborhoods (Han 2019, 186). The same pressures face a group of local residents in Dublin who have collectively acquired buildings to facilitate affordable "independent spaces" for work, art, or socializing, which are "frequented by a wide variety of people, from trendy artists to asylum seekers, from working class ravers to anarcho-punks, and from community activists to isolated young people and those with mental health difficulties" (Bresnihan and Byrne 2015, 8). These spaces, acquired in part as a response to rapid urban development, increased rents, the commercialization of street life, and the privatization of public space, are now vulnerable to those same forces.

As a response to gentrification pressures, urban communities and "commoners" are pushing to transform the legal status of the land and buildings they utilize and/or occupy to place them under community control through legal and property mechanisms such as limited--equity cooperatives (LECs) or community land trusts (CLTs), which are mentioned in chapter 1. LECs and CLTs are designed to allow communities to self-govern and steward urban land and buildings and to keep them affordable and accessible to future users. Land and buildings managed and governed as an LEC or CLT are dedicated to low-cost housing and commercial spaces as well as urban farming or community gardens. LECs, for instance, differ from traditional housing cooperatives in that they ensure long-term affordability by removing the housing from the speculative market, limiting the resale amount, and collectively subsidizing low-income owners.

CLTs operate most uniquely as a steward of shared resources by removing land from the speculative market and separating land *ownership* from land *use*. The original CLT was created to be "a legal entity, a quasi-public body, chartered to hold land in stewardship for all mankind present and future while protecting the legitimate use-rights of its residents" (Swann et al. 1972, 10). The CLT operates as a nonprofit entity that holds legal title to the land and enters into long-term "ground leases" with those who utilize the land for apartments, homes, commercial buildings, or green space. In some cases, instead of ground leases, land users receive a warranty or surface rights deed that secures their right to use the land and to pass it along to their heirs. However, the CLT maintains ownership of the land underneath any building or structure and thus controls the future use of that land, including its affordability. Individual users own the buildings or structure on top of the land and enjoy all the benefits of that ownership—including using, improving, excluding others from it, and mortgaging it. The buildings on top of land can also be transferred or sold by users for an amount determined by the resale formula set forth in the ground lease or deed, allowing a small profit to be made from the sale but otherwise keeping the land affordable for the future purchasers. The CLT may also retain a first right of refusal to purchase the building or unit whenever it is being sold. The terms of the ground lease and all other conditions of land ownership/use are set by a tripartite board of directors, which governs the CLT.

CLTs and LECs resonate with what property scholars refer to as *governance property*. Governance property characterizes many (if not most) forms of private property ownership today in that such property is shared with multiple owners or users collectively making decisions and rules about access, use, enjoyment, and transfer of property (Alexander 2012). Governance property is a departure from the prevailing property ownership model, characteristic of Western legal culture, which aggregates all legal rights and entitlements in one owner. As property scholars have begun to recognize, the dominant Western model of property and resource ownership—the *fee simple*—looks more and more ill-fitting for the urbanized, interdependent world in which most people live. Endowing owners (public or private) with a monopoly on urban land and resources, this form of legal ownership "misses most of how urban property creates value" through spatial relationships that result from the density and proximity characteristic of urbanization (Fennell 2016, 1460–61).

To meet the demands of contemporary urban land use requires instead a mix of approaches to mediate access to resources, particularly for those who have much less access to them. It requires, at the very least, embracing approaches that recognize relational property interests and resource governance in ways that advance access to urban resources for the most vulnerable and marginalized communities facing resource uncertainty and precarity. More to the point, it is possible to adapt and unbundle the legal entitlements to access and use property to fit the normative aims of the commons that satisfies various commitments to social inclusion and distributive justice (Marella 2017). Those legal entitlements can be re-allocated to different owners or users and/or limited through legal restrictions that make possible the inclusion of different classes of rights holders. In other words, the bundle of legal entitlements or rights need not be aggregated in one owner (or even a collection of owners) and need not be without internal limits or restriction.

LECs and CLTs place internal limits on the right to hold and sell property, limits that go against the rights that owners would have in traditional private property arrangements in which the owner or owners have total freedom in regard to how to use, sell, or transfer property subject only to external constraints such as zoning or environmental regulations. In return, limited-equity owners gain sustainable wealth building opportunities and lasting affordability. CLTs are governed collaboratively by the users of the property, typically low-income residents, along with the larger community and representatives from government and often the private sector to construct and sustain the buildings, infrastructure, and maintenance over the long term. The traditional governing board of a CLT is *tripartite*—that is, it comprises an equal number of seats represented by users or people who lease the land from the CLT, residents from the surrounding community who do not lease land from the CLT, and the public, who are represented by a variety of stakeholders such as public officials, local funders, and nonprofit providers of housing or social services. CLTs are rooted in a desire to exercise community control over land, to remove land from the speculative market, and to facilitate sustainable uses that benefit disadvantaged communities (DeFilippis et al. 2018).

The traditional governance structure of the typical CLT notably differs from the kind of closed, private governance of condos, co-ops, and other *common interest communities*. LECs, by contrast, are in many ways akin to common interest communities like condominiums and traditional co-ops. A cooperative is governed by a board but consists exclusively of private property owners. Unlike in a traditional co-op, however, in an LEC the owners can restrict the resale and equity gains to keep the housing affordable. They do so by private agreement among private property owners. Those owners can change this agreement at any time to make private gains from speculation, as occurred in the hundreds of cooperative agreements in NYC that converted to market rate units. CLTs, on the other hand, transform what might otherwise be a collection of individuals owning property, as in a traditional housing cooperative, into a collaboratively governed nonprofit institution that creates a form of collective ownership for the common good through its democratic governance structure.

Both CLTs and LECs are conceived as a way to ensure that critical urban resources remain accessible to individuals and communities by adapting private property entitlements to the norms of a common good. They maintain affordability and hence accessibility of the resource by limiting the amount of equity that can be extracted from these goods so that future generations can share in their use and by creating stable property rights for those who occupy and use the good; they accomplish this through a governance structure that maintains control over the good or service within the community served. As Lisa Alexander has written, property stewardship is created by removing the profit motive and by allocating rights and responsibilities in a way that gives stewards decision-making control over resources in a manner similar to ownership but without the emphasis on sole dominion and the individual exchange value of property (Alexander 2019, 402). In other words, stewardship grants control of and access to resources without formal fee simple title and without wealth maximization as a goal of property access, and it "connects stewards to economic resources and social networks that maximize their self-actualization, privacy, human flourishing, and community participation" (Alexander 2019, 404).

Stewardship encourages co-management, co-development, and construction for the common good. It is not antidevelopment nor antiwealth-building. It discourages economic development in the absence of community building. It privileges the right to be included in community over the right to exclude from collective resources. It favors collective, community wealth building over individual wealth maximization, although it can create a path for both. Consider a recent 2019 study of fifty-eight shared-equity homeownership programs and 4,108 properties over the past three decades in the US that analyzed the characteristics of households owning shared-equity homes and the performance of these forms of property ownership across the nation (Wang et al. 2019). The study focused on the three most common types of these programs—CLTs, LECs, and deed-restricted homes. The study found that limited-equity models of homeownership serve predominantly first-time homeowners who tend to be members of vulnerable populations, particularly low-income racial and ethnic minorities, and female-headed households. Limited equity homeownership not only provides stable and affordable housing across generations, but it also provides for financial security and mitigates risk during times of economic turmoil (e.g., fewer home foreclosures). More specifically the study found that this form of homeownership can be a pathway to entry to the larger market for homeowners. Six of ten limited-equity homeowners used their earned (though limited) equity to eventually purchase a traditional market-rate home.

To appreciate the potential of CLTs as a stewardship model, a closer look at the famous Dudley Street experiment, discussed in chapter 1, is instructive. Dudley Street, located in Roxbury, Massachusetts, was one of the most economically distressed neighborhoods in the Boston metropolitan region in the late 1980s and early 1990s. After cleaning up many of the vacant lots that littered its neighborhood (a mix of city-owned and privately owned parcels), Dudley Street residents incorporated as a nonprofit (DSNI) and embarked on an ambitious plan to create an "urban village" that would develop the neighborhood without resulting in any displacement of the existing residents (Medoff and Sklar 1994). To do this, DSNI, along with its community partners, approached the Boston Redevelopment Authority and requested eminent domain authority, which was granted by the city of Boston with the support of the newly elected mayor of Boston, Ray Flynn. With this authority, the DSNI assumed control of over 1,300 vacant parcels and created a community land trust, Dudley Neighbors, Inc., that would own and secure that land for long-term affordability.

The once vacant land has been transformed into more than 225 new affordable homes, a 10,000 square foot community greenhouse on the site of a former auto body shop, two acres of community farms, playgrounds, gardens (that today total more than seventy), commercial space, and other amenities of a thriving urban village (Smith and Hernandez 2020, 288). The housing now includes ninety-seven homeownership units, seventy-seven limited-equity cooperative units, fifty-five rental apartments, and ninety-six individually owned homes. Consistent with its neighborhood plan, the majority of housing units are targeted for families making between 30 and 60 percent of the area's median income, which equals approximately $30,000–$60,000 for a family of four. Individuals or families who wish to purchase one of Dudley's affordable homes participate in a lottery system in order to ensure equal and fair access to the homes that become available. After a purchase, the homeowner pays a small lease fee for the land on which the house sits; the land continues to be owned by the CLT. The homeowner also agrees that if the home is ever sold, which is rare in the Dudley area, the home must be sold at a cost determined by the formula used by DSNI's CLT. The sustainability of the Dudley model has been proven in part by the fact that during the economic crisis in the period 2008–2013, there were no foreclosures of DNI homes even as the surrounding neighborhood had more than two hundred foreclosures (Smith and Hernandez 2020, 290).

The DSNI CLT is democratically governed, organized, and run so that each cultural-ethnic grouping present in the Dudley community gets an equal voice. The Board has thirty-five seats, twenty of which are reserved for community residents including an equal number of representatives of the four main ethnic groups inside the community. Of the twenty community seats, four seats are for Black residents, four are for Latinos, four are for residents with a Cape Verdean heritage, four are for white residents, and four are for youth (ages 15–18) living in the community. Of the remaining seats, two are for community development organizations, two for local religious organizations, seven for partner organizations, and two for small businesses in the community. Once in place, these thirty-three members then elect two additional members from those who wanted to participate on the board but were not elected, for a total of thirty-five. Residents alone vote on who gets to serve on the two-year board term. Campaigns are door to door and face to face so that all residents have the opportunity to meet the members of their board. Once elected, the board approves all decisions made by DSNI. All projects and campaigns must be vetted and approved by the board, but such decisions are always open to community input and participation.

The Caño Martín Peña CLT in Puerto Rico, which includes more than 270 acres of land across eight neighborhoods in an informal settlement, has a slightly different governance structure that is smaller but equally as representative of the collective interests in the land stewarded by the CLT. The CLT was enabled by the Public Authority, which facilitated the financing of the project, and supported by the creation of the Martín Peña Canal Special Planning District, a district of two hundred acres of public land transferred to the project for the creation and management of a CLT. The governance structure of the Caño CLT was collectively decided upon by representatives from the eight Martin Peña neighborhoods in a participatory process that resulted in a local regulation that established the legal basis for the CLT, including its governance, rules, and procedures, and community stewardship of the land in perpetuity (Hernández-Torrales et al. 2020, 198–199).

The local law provides that at least 45 percent of the CLT board must be composed of the district's own residents, giving the communities a strong governing role. The eleven-member board of trustees consists of representatives of CLT users, the larger community, private entities, and state and local government. Of the eleven members, four are residents whose homes are located on the land owned by the CLT (elected by an assembly of trustees), two are residents of the surrounding communities (designated to serve by the eight organizations that formed the CLT), two are nonresidents of the district and selected by CLT board members on that basis of the skills and knowledge they can contribute to the CLT, and the remaining three seats are occupied by representatives of state and local government consisting of appointees from the local development corporation, the mayor of San Juan, and the governor of Puerto Rico (Algoed et al. 2018).

CLT members are also in charge of spreading the concepts of "collective ownership" and work closely with professionals, professors, experts, and students and are invited to take part in workshops and meetings in order to share their experiences (World Habitat 2015; Bernardi 2017a). The CLT also supports residents with financial education and specific programs to promote citizens' participation and critical awareness in order to address and improve social justice, affordable housing, food security, violence prevention, youth leadership, adult literacy, and local entrepreneurship (Bernardi 2017).

NESTED URBAN COMMONS

One question raised by CLTs is whether and how much they can *scale* from a site or location to a citywide or even regionwide network of stewarded and co-governed land and resources. Recently, we have begun to see that they can scale through networking and with the support of local policies and public resources. The use of community land trusts to protect and sustain access to affordable urban goods such as housing, commercial space, and green resources is expanding to protect these resources at the neighborhood level and even across an entire city and region. For instance, building on its successful model, Dudley Street Neighbors and ten other neighborhood groups from across Boston in 2015 launched the Greater Boston Community Land Trust Network to expand the CLT model even at a time when acquisition of urban land has been made more difficult because of rising land values and rapidly gentrifying cities. This network has supported and seen the rise of five new CLTs—the Chinatown CLT, Somerville CLT, Boston Neighborhood CLT, and Urban Farming CLT—across the Boston metropolitan area, and it is beginning to push for municipal policies and public resources to support their expansion and growth (Smith and Hernandez 2020, 294).

The citywide NeighborSpace land trust in Chicago is another example this kind of scaling. NeighborSpace is an independent, nonprofit land trust that preserves urban land throughout the city of Chicago for community gardens and open space. Created in 1996 by three government entities—the city of Chicago, the Chicago Park District, and the Forest Preserve District of Cook County—NeighborSpace now oversees a water-based project and 129 land-based sites located in thirty-three wards across the city, many of which are involved in community gardening projects. NeighborSpace's primary goal is to preserve and protect community-managed open spaces, particularly in areas where open space is lacking or vanishing, which tends to be the case in underserved areas. The idea for NeighborSpace grew out of the city space plan. City leaders became increasingly concerned about the lack of open space in Chicago and the vanishing number of vacant plots being bought by private developers. In 1994, a consortium of the three government entities named previously brought together community leaders, residents, and nonprofit organizations to brainstorm possible solutions to this ever-growing problem. From these efforts, they created NeighborSpace, inspired by a recognition that many community members were, on an informal and ad hoc basis, already working together to revive, enjoy, and preserve vacant or blighted land in their communities. In an example of social and economic pooling, NeighborSpace continues to receive the active support of the city government, the broader Chicago community, the many foundations and philanthropists that provide donations, and the teams of gardeners, composters, and other community actors that perform the day-to-day work on the land. Because of such support, NeighborSpace continues to grow in scope and impact; indeed, each year, it acquires between three and five new land plots, on behalf of dedicated community members and groups that maintain them.

NeighborSpace is unique among land trusts because it represents the kind of nested and multilevel governance structure, as previously mentioned in chapter 1, that Ostrom and others have found can constitute a *polycentric* system of governance that can be more effective than a monocentric system in efficiently delivering local goods and services. Ostrom's work on user-managed common pool resources similarly found that, for large scale resources, higher level public authorities played an essential role in supporting resource users in the management of these resources. NeighborSpace is thus a separate public enterprise operating as a sublocal layer of governance "nested" within the county and city government, and which is supported by those higher levels of government. Managing shared resources at a complex (in this case, citywide) scale can involve self-organized small units or groups of users acting relatively autonomously but within a federated system that links them together.

Once a land grant is established, NeighborSpace generally relinquishes operational control to the land trust, which transfers most of that control to the local gardeners and community groups that act as stewards over the land. In effect, NeighborSpace operates as a higher-level authority, whereas the real control and management over day-to-day affairs is handled by local members and groups in the community in which the land is located. NeighborSpace, the land trust, handles the land purchases; performs environmental assessments and title work; holds the titles, easements, or leases that it acquires; provides liability insurance and legal defense; and works to secure a dedicated water line for every parcel of land that it obtains. It also provides some guidance and other forms of support, "including a signage template, a list of gardeners' rights and responsibilities, and a tool lending library," and it acts as the liaison between the government and the participating community groups.

However, NeighborSpace is not involved in the day-to-day management of the land plots, which is left to the community, and plot users, in what is described as a "nonhierarchical" governance structure that prevents the centralization of power in any one individual's (or one group's) hands. Although most of the gardens are not "allotment" gardens, in that they don't always have plot holders (both rather plot users), communities nevertheless have created an array of garden types, ranging from vegetable gardens to riparian habitats. For whatever type of garden is created, the rules of the land trust require collective governance over the acquired plots. Moreover, governance rules prohibit a single lead gardener or overseer and instead require multiple leaders overseeing the land's development, as well as community support and buy-in.

NeighborSpace is unique among land trusts in another way. It is distinguishable, for example, from the kind of *community* land trusts previously mentioned in this chapter and chapter 1. The land managed by NeighborSpace is a mix of land owned by the trust itself and that owned by other governmental and private entities. Approximately 20 percent of the lots managed by NeighborSpace are leased or utilized with permission from government units and private entities that maintain ownership of that land. This raises another concern expressed by some in regard to what is lost when community-stewarded resources are managed at a larger scale, such as a city-wide urban land trust in the case of NeighborSpace. There are similar land trusts emerging at the city-level, most often focused on developing affordable housing on a city-wide scale, which are controlled almost exclusively by local public officials. Some, like the Atlanta Land Trust, created to maintain affordability in neighborhoods at risk of gentrification and displacement, emerged after an attempt to establish a community land trust fell short for lack of resources and capacity to organize the community. Establishing the land bank as a more public-oriented, versus community-oriented, entity can bring crucial funding and support. On the other hand, some bemoan the loss of "community" in these land trusts and the loss of community control of land and resources (DeFilippis et al. 2018).

As some have noted, the traditional "tripartite governance" governance structure is not always followed when new land trusts are created in cities (Miller 2013, 5). Many newer city-wide land trusts and community land trusts may be falling prey to the danger of "capture" in which land use and development decisions are made by a group of elites that invest in projects serving a narrow set of interests and which do not always align with the community's desire or needs. As DeFilippis et al. note, the "community" part of CLTs was originally conceptualized as a nested set of relations, involving a resident community living in the trust, a wider community of residents and others who represent the broader community that would benefit from a restructuring of land ownership practices. Today, however, some CLTs have moved away from these overarching aims and have simply become a tool to provide affordable individual homeownership in expensive markets. The focus on individual ownership, however, detracts from the attention on using CLTs to empower disadvantaged communities and for community control of land and neighborhoods (DeFilippis et al. 2018).

Some of these drawbacks, however, may have less to do with scale than with the conditions under which a distributed, polycentric system of co-governance can flourish. In other words, the design of well-structured governance institutions at various scales can prevent some of these drawbacks through clear governance rules to ensure strong representation from the most vulnerable stakeholders, procedures for entry and exit of governing stakeholders, and establishing clear normative values (such as permanent affordability and community control) that guide these institutions. For this reason, the Dudley Street CLT has been held up as a model of what legal scholar Anna di Robilant has called "democratic deliberative" property (di Robilant 2014). This form of governance property promotes its public-oriented character through decision making, enforcement, and monitoring by "multiple actors affected by the use of resources that implicate public values and collective interests" (di Robilant 2014, 306). The idea of democratic deliberative property maps nicely onto the notion of resource stewardship in that "decisions concerning the use and management of resources that implicate fundamental public interests" are not made by a single owner, even if the owner is a public official or agency, but rather "through a more deliberative democratic process in which representatives of affected parties participate as equals and give one another reasons that are mutually acceptable" (di Robilant 2014, 304–305). The challenge, however, is how to manage the design or co-design processes so that the dangers and risks attendant to allowing these autonomous institutions to flourish and govern common resources reflect deliberative democratic values and accountability to the communities that they are set up to benefit.

What stabilizes the kind of distributed co-governance ecosystem that we envision in the next few chapters is the role of the public authority, which become the enabler and facilitator of the creation and maintenance of urban commons and ultimately of political and economic redistribution. In the next chapter, we turn to the emergence of city policies that enable, facilitate, and support urban commons and allow them to nest within the governance infrastructure of the city. Whereas communities and other stakeholders organize themselves autonomously as potential partners that can collectively manage urban resources, city officials and staff are tasked to assist, collaborate, and provide technical guidance (such as data, legal advice, communication strategy, design strategies, and sustainability models) to those efforts. The governance output that emerges from implementation of these policies is the co-design of a variety of urban commons as well as the co-production of community goods and services at the city and neighborhood level. These very sophisticated processes and institutional architectures are new and complex to design, as we discuss in the following chapters, and they do not always function as they should in an ideal world. However, these policies are windows into an alternative vision of city governance in which heterogeneous individuals and institutions can come together to co-create or co-govern the city, or parts of the city, as a commons.

3
THE CITY AS A COMMONS

In chapter 2 we offered examples of collectively or collaboratively managed urban resources, ranging from community gardens and parks to housing and commercial spaces. At a small-scale level such as a community garden or urban park, users come together to build and manage these resources, often with the support of the local government. This state enabling includes tacit or express consent to utilize public land or resources, the transfer of resources (financial, technical, or other), and changes in local laws or practices. For large-scale urban resources such as an urban village or a neighborhood, collectively managed resources emerge and are sustained with public and private resources and an institutional structure, such as a community land trust, that matches the scale of the resource. These institutional structures can be nested in a larger (local or regional) governance system that allows users autonomy over the resources without subsuming them into a centralized governance regime. The question is whether these examples represent interesting ad hoc experimentations or remain largely disconnected from the normal operation and governance of cities. Alternatively, is it possible to put in place policies, institutional practices, and social infrastructure? To root and encourage these experimentations across a city? In other words, are there policies or practices that enable a city, or parts of a city, to operate as urban commons?

In this chapter, we examine the emergence of public policies in a handful of cities that enable public and private actors to collectively or collaboratively create and then steward shared urban resources throughout the city. This chapter offers a largely descriptive analysis of efforts by many cities to establish a kind of *urban co-governance* regime, a concept that we unpack more critically in the next chapter.

Collaboration, as a general principle, has emerged in governance studies to replace adversarial and managerial modes of policy making and implementation (Ansell and Gash 2008). In this model, several stakeholders interact to implement public policies or manage crucial assets for the community. In order to accomplish collaborative policymaking and community asset management, relevant stakeholders, consisting of individuals and groups that have an interest in a process or outcome, must engage with one another to address issues that cannot easily be solved by any one of them on their own (Bingham 2009, 274). Collaboration is also a mechanism that is central to the governance of common pool resources, that is, shared resources on which discrete and numerous individuals rely. These common pool resources are unique in that they create a relationship of interdependence among the public officials and the public at large, specifically communities in which these resources are located. As such, when we refer to the *public*, we want to acknowledge that there are two forms of public actors: the public conceived of as the public sector (the city government and the municipal administration or bureaucracy) and the public conceived of as the general community of city residents or a specific community of residents. Collaboration is a methodological tool that enables these two kinds of public actors to work with each other and/or to work with private actors that exist outside of these two publics. As the policies described in this chapter attest, cities can utilize collaboration as a methodological tool through which heterogeneous individuals and institutions co-create or co-govern the city or parts of the city as a common resource.

The kind of collaborative governance reflected in the city policies described in this chapter endeavors to deeply engage citizens through public-public and public-community partnerships with the goal of implementing an arrangement in which citizens are governing the city and are not simply being governed. The policies described here treat at least part of the infrastructure of the city as a common pool resource available for residents and others to come together over and to collaboratively utilize for the provision and delivery of essential goods and services. This infrastructure can range from vacant land to underutilized buildings to digital networks. The policies adopted by these cities allow access to this infrastructure so that residents can, by their own initiative and supported by public (and increasingly also private) resources, directly take action to address a range of challenges confronted by a community or neighborhood. The positive effects of these actions are not limited to the realization of small-scale projects but also have taken forms such as the revitalization of degraded urban spaces. At the heart of the concept of the city as a commons is the idea that urban resources should be accessible to and shared more widely with communities that are lacking in the resources necessary to survive and thrive in cities. As we have argued, reconceiving the city as a commons can be a powerful tool to fight inequality in cities. Policies such as those described in this chapter are motivated, at least in part, by the desire to more broadly share urban assets and resources that can in turn be used to more fairly distribute social and economic wealth.

We are mindful, of course, of the tendency to romanticize collaboration in urban planning processes. Too many experiences of failed collaborative practices simply devolve planning processes to the sublocal level without offering new tools and resources to enable meaningful collaboration, including increasing the capacity of individual actors and vulnerable communities (Elwood 2002, 2004). The best collaborative urban development processes, in our view, deeply engage and empower a wide range of actors in the revitalization of city spaces and in the management of city assets. For each attempt to grant or recognize the right to collectively steward or own shared urban resources we would pose the question whether the collaborative process leading to these ends are truly effective at engaging underrepresented groups and marginalized communities, given the history of dispossession and exclusion in so many contexts. We have observed that the answer often depends on the consciousness and intentional efforts of those designing these arenas and processes. On the basis of our experience, we believe that it is necessary to constantly interrogate policies like those we describe here and to heed the lessons of failed collaborative urban governance practices.

The policies described in this chapter situate the local government as an enabler and facilitator of collaboration and ultimately of political and economic redistribution through shared urban goods and infrastructure. If the city itself is a shared resource, then a strong collaborative system of decision making should also nudge toward redistributing some of the assets of the city to support differently situated individuals and communities within the city. This idea is akin to the "city-making" that local government scholar Gerald Frug (2001) proposed, in which he advocated transforming cities and city services into vehicles for community building across local government boundaries. In a similar way, a commons-based governance approach envisions cities as vehicles for collaboration across formal and informal institutional arrangements. In the following chapters, we show that this facilitative function needs to be somehow embedded in a formal act (e.g., official policies, regulations, contracts, agreements and other legal instruments that allow the cooperation between actors with different stakes in the process or that hold different skills and resources) and adapted to the local context as well as situated at the most appropriate scale (e.g., at city or at neighborhood level). Sometimes this function can even be exercised by other agents in the city. Collaboration implies that some agency or some institution needs to act as an institutional platform enabling multistakeholder cooperation in the city. This agent or agents should be trusted enough to undertake the task of identifying common challenges or needs to match potential collaborators, to help call them to the table, to facilitate the sharing of knowledge and experiences, and to monitor and measure the development and impact of the cooperation.

The city government, of course, can serve as an enabler or pivotal agent. The modern city has traditionally been envisioned as the Gargantua: "a single metropolitan government or at least the establishment of a regional superstructure which points in that direction" (Wood 1958, 122). According to this model of the city as a centralized agency that annexes new territory to encompass all residents of a region, the government establishes and manages citywide services such as transit and utilities. It acts as a single unit and, as the principal decision-making structure, should "best able to deal with metropolitan-wide problems at the metropolitan level" (Ostrom, Tiebout, and Warren 1961, 837). Notably, however, as Gargantua's power and capabilities grow, so does the "complexity of its own hierarchical or bureaucratic structure" (Ostrom, Tiebout, and Warren 1961, 837). This decreases its efficiency and increases the likelihood of failure in reacting to local issues. Thus, the megacity model of governance is plagued with internal difficulties.

The idea of the public authority as a facilitator—a relational state (Cook and Muir 2012)—is part of the move from a "command and control" system of government to what we have previously called "urban collaborative governance" (Foster and Iaione 2016), a system that at its core redistributes decision-making power and influence away from the center and toward an engaged public. The facilitator state creates the conditions under which citizens can develop collaborative relationships with each other and cooperate both together and with public authorities to take care of common resources, including the city itself as a resource (Grafton 2000). The facilitator state can also be understood as part of the move from government to *governance* in urban and global politics. Urban governance includes not only the traditional, hierarchical forms of decision making by public actors but also the influence of other nongovernmental actors and sectors (e.g., the media, NGOs, investors and donors, industry, and knowledge institutions) on those actors and their decisions. As the policies described next reflect, urban governance often involves contestation, negotiation, and compromise about the allocation of political power and material resources.

THE DECLARATORY VERSUS CONSTITUTIVE APPROACH

To better understand these policies, we have organized them into two categories according to their functional approach to the creation and sustenance of collectively governed urban resources in communities, called urban commons. The first category we call a *declaratory* policy or law. The second category we call a *constitutive* policy or law.

The declaratory policy acknowledges the existence of collectively managed individual resources or neighborhood institutions as forms of urban commons. Pursuant to a declaratory policy, local governments officially recognize the right of these communities to emerge and self-organize, and they acknowledge and even promote their collective governance practices or institutional form. This might entail recognition of social norms agreed upon by the community (e.g., in the case of Naples discussed further on) and/or validation of the public value produced by the community that justifies their right to utilize the shared resource (i.e., in the case of Barcelona). The local government might even enter into some sort of agreement with the collective, lending legitimacy and some stability to the effort as well as indirectly encouraging other bottom-up efforts throughout the city.

The second category, constitutive policies, includes those policies that embody a more top-down, institutionalized approach. Policies that have a constitutive effect create in favor of or grant communities rights to govern assets, infrastructure, services in the city and therefore are specifically aimed at encouraging the creation of urban commons throughout the city. They can create the conditions for governing some city resources collaboratively by offering new legal authority or adapting existing laws. The legal tools provided by these city policies include newly created authority for collaboration pacts between local government and a designated community or group (e.g., in the case of Bologna and Turin in Italy) or agreements pursuant to an existing legal framework (e.g., in the case of Madrid, Spain). When the city signs such pacts or agreements with city residents (who are either formally organized into a legal entity such as an NGO or are entering the pact/agreement as informal groups) they aim at constituting co-governance arrangements to enable the emergence of urban commons throughout the city.

Both approaches present ongoing challenges and attendant costs. City policies embodying a constitutive approach, for instance, can be a tool for empowering urban communities or for controlling and manipulating them. Critics have rightly noted that these policies are too susceptible to become yet another vehicle for public-private partnerships (PPPs). Although PPPs arguably avoid the inefficiencies of a top-down, command-and-control system, they are subject to various problems, including cutting out what we refer to as the second *public*—everyday citizens and communities. On the other hand, it is also true that these policies need not mimic PPPs nor embody their costs. As we observe further on, in many cases these policies evolve and develop in ways that represent a first step in or instigation of deeper collaborative processes that activate and engage residents and communities in collectively utilizing their shared resources toward particular social ends and solving conflicts over these same resources.

In other words, collaboration pacts represent one tool for shared and even bilateral governance with multiple actors or between two types of actors, whether public-private or public institutions–public citizens. In this evolutionary process, we can observe a transition from a city characterized by a vertically oriented governance structure to a horizontally organized structure that seeds institutional spaces in which citizens operate as peer co-workers and co-designers; addresses social and economic conflicts together; and resists urban shrinkage or gentrification processes, two opposite phenomena that can both lead to negative consequences. On one hand, there are neighborhoods abandoned by their inhabitants because, for instance, they lack job opportunities. On the other hand, low-income populations can be forced to leave certain neighborhoods because of increased housing and living costs (Richardson and Nam 2014). By working together, collectivities of actors can contribute to counteract these challenges and foster inclusive and equitable city-making. The institutional settings in which urban collaborative governance can operate are places of networking and of connecting and coordinating different and autonomous actions for the same shared goals.

Similarly, city policies embodying a declaratory approach have costs and challenges. For one, they can expose the municipality that wants to legitimize self-organization practices and stewardship of local resources to the accusation that they are regularizing illegal occupations or privatizing public assets through social collectives. These collectives could manipulate political leaders in exchange for electoral support or otherwise exert political pressure to wrest control over available, unclaimed local resources. Moreover, legitimizing small collectives or groups to manage shared resources, so-called urban informality practices, can risk becoming exclusionary. Even though most of these practices start as open and inclusive, embedding these principles in an institutional design, they tend to become very homogeneous over time, with members tied together by similar political beliefs, ideologies, social norms, and values. Perhaps tight-knit groups, especially those with fewer resources, should be able to control resources, and they have done so successfully, as Ostrom found. However, we should at least approach homogeneity with a critical eye, considering understandable concerns about interest-group factionalism, capture, and even illegal use of force by small factions. Recognizing these social practices without setting forth precise criteria and monitoring mechanisms could end up giving a free pass to control urban assets and resources and giving leverage to social groups tainted by racial or ethnic bias under the guise of recovering urban assets and resources for the common good.

The power and challenge of both the declaratory and constitutive approach lie in the fact that they represent decentralized governance. Their structure resembles a "loosely coupled system," subject to fraying at the margins and not glued together enough to be organizationally coherent (Weick 1976). Ironically, it is the central (local) government that can play a stabilizing role, which becomes that of coordinator and facilitator by providing spaces and platforms for collaborative and co-design processes. These new collective or collaborative arenas are user-centered open spaces that aggregate different actors (including city inhabitants, social organizations, private enterprises, public institutions, and knowledge institutions) in participatory and co-design processes in which decisions could be taken or new public policies can be generated outside of the local government. In this sense, the networks, actions, and reactions of others in a collaborative ecosystem are independent and autonomous but nested within the local government, consistent with the kind of polycentric system we mentioned in previous chapters. Elected officials no longer behave as citizens' representatives but rather as collaborative institutional managers. City officials and staff are tasked with assisting, supporting, collaborating, and providing tools and guidance (such as data, legal advice, communication strategy, design strategies, and sustainability models) to enable themselves to manage, mediate, and coordinate the ecosystem. The role of a public official is therefore that of manager: enabling and supporting (and perhaps coordinating) parts of the ecosystem to allow it to *nest* within the larger policy of the city.

The policies that we describe in this chapter and that represent an attempt to share urban assets and resources more broadly within a city introduce a heightened level of institutional and governance complexity. More than networking actors and communities governing shared urban assets, urban co-governance ultimately entails a different type of institutional complexity. This requires a symbolic shift rather than a change of actors or networks in power (Lievens 2014, 14). In other words, in an urban collective and collaborative democracy, governance needs an institutional platform from which the politics can become visible, equal, contestable, and legitimate. These are platforms by which the relationship between power, law, and knowledge is redefined. They are places from which instead of hierarchies of power and wildly unequal bargaining positions, we see networks of empowered inhabitants and stakeholders self-organizing, co-creating, co-designing, and co-implementing planning and policy solutions for complex urban environments together with policymakers and local officials. To accomplish these ends, collaboration cannot be centralized within the Gargantua itself but rather needs to be distributed around the city and specifically to the places and populations that are the most marginalized and disempowered. We show that in the iteration and evolution of local policies and regulations described in this chapter, they eventually lead to the dispersion of the spaces of collaboration away from the center, throughout the city and to those places that represent the margins of social and economic life.

THE CONSTITUTIVE APPROACH: ENABLING COLLABORATION AND SHARING OF CITY RESOURCES

Many of the early city policies described in this chapter embrace a regulatory tool—*collaboration pacts* or their equivalents—as a vehicle for collective action by local citizens to engage in shared use and management of urban resources. Collaboration pacts exemplify the constitutive approach to treating the city's infrastructure, its assets, and public resources as a shared commons. Collaboration pacts were first adopted by the city of Bologna; in them, city residents, social innovators, entrepreneurs, civil society organizations, and knowledge institutions, along with the city, could agree to co-design governance of various urban goods and services. Although broadly conceived of as a regulatory tool, the impetus for these policies was to anchor the public—both the public sector and the unorganized public—as the main actors in the collaborative or collective enterprise and then to bring in other actors or sectors. Bologna's experience was a catalyst for other cities (particularly in Italy but more broadly in Europe) to think about how to reshape the relationship between citizens and the local administration in relation to the use of urban resources and delivery of urban services. We describe here the main experimentations in collaborative city making that were inspired by and/or followed the path of the Bologna experiment. Reflecting on these experimentations and taking what we know from their evaluation has revealed sufficient reasons for skepticism about their practical implementation.

Before describing and analyzing the Bologna experiment, it is worth noting an important precursor to Bologna which also represents an early constitutive approach to treating the city as a commons.

The city of Seoul (Seoul Metropolitan Government, South Korea) in 2012 was the first city in the world to build a policy program to promote the sharing and collaborative economy at the city and district level through a multiyear policy program, the Seoul Sharing city policy, which kicked off in 2011 and is still ongoing. The program's legal basis is the Seoul Metropolitan Government Ordinance on the Promotion of Sharing (hereinafter called the Seoul Ordinance or simply Ordinance).[1] The Seoul Sharing city policy is an important forerunner of the urban co-governance approach (Won-Soon 2014). Seoul is the capital and largest metropolis of the Republic of Korea (commonly known as South Korea). Together with the neighboring Incheon metropolis and Gyeonggi province it forms the Seoul Capital Area, which houses up to half the country's population of 50.22 million people, with 678,102 international residents and a GDP of $1,005,538, accounting for 49.0 percent of the national GDP (UN-Habitat 2020). Seoul is a leading and rising global city and has experienced dramatic economic growth that has transformed it into a competitive economy. However, inequality is on the rise. As with other global cities, Seoul is facing a challenge from the choice between being a competitive global city and embracing balanced development (Kim and Han 2012).

The Seoul Sharing city policy implementation can be outlined in three phases: phase 1 (2012–2014), the foundational phase; phase 2 (2015–2019), focused on expanding the users who enjoy access to the opportunities of the sharing economy and on consolidating the international standing of Seoul as a city promoting the sharing economy; and phase 3 (2021–ongoing), during which the city is catalyzing serious efforts to turn city residents from mere receivers of the public support and subsidies— that is, users of the sharing economy—to proactive actors and is building a resident-led public-private partnership to be the groundwork for the sharing economy to thrive alongside principles of ecological sustainability, inclusiveness, and civic autonomy.

Phase 1 was the foundational phase. It was focused mostly on enacting the Ordinance at the districts/city level, supporting small companies oriented toward the sharing economy. The goal of this first phase was increasing the citizens' awareness of the sharing economy, activating as many sharing-based services as possible, and supporting sharing companies. This phase emphasized the role of suppliers in nurturing the sharing economy.

In 2012, the Seoul Metropolitan Government under the mayorship of the late mayor Park Won-Soon, who had been a social justice and human rights activist before becoming mayor, first declared its intention of becoming a sharing city. Shortly after, the city enacted the Ordinance designating sharing organizations and enterprises, providing a Sharing Promotion Fund, and organizing sharing schools and communication activities (Foster and Iaione 2016; Bernardi 2018). Through the Ordinance, the government supported the creation of new sharing businesses and start-ups, including the sharing of car, bikes, children's school clothes (Kiple n.d.), car parking (Johnson 2014), and meals (Seoul Share Hub 2016)—the last with the purpose of establishing a social dining practice. The city supported the sharing of vacant public spaces through Seoul Ordinance §5323 on the "opening and use of dormant spaces and public facilities," passed Dec. 31, 2012 (Seoul, South Korea 2012). Another important service promoted in the first phase is the sharing of housing (Guerrini 2014) with a program based on creating the connection between elders with spare rooms and students, to promote intergenerational solidarity and collaboration, called "Same roof generation sympathy." In addition, to encourage entrepreneurialism in the education of young residents, the government organized sharing schools and startup schools (Johnson 2014). Finally, the sharing of data and digital services was encouraged from the very beginning with the creation of city-led public data sharing as well as public Wi-Fi.

The policies have generated a huge number of initiatives, seen massive participation by local residents, garnered several international awards, and inspired the creation of a Sharing Economy International Advisory Board composed of international scholars. An important role in the policies' success is the establishment of the Seoul Community Support Center (hereinafter called SCSC), a network mediator to encourage grassroots participation in neighborhood initiatives and encourage a sense of local identity and belonging in Seoul's urban communities.

The city also lends technical and other support to sharing companies, including public relations and training. The Seoul policy also evinces a redistributive purpose. To solve potential obstacles to sharing for some populations, such as those due to the digital divide, the city offers free Wi-Fi inside the subway, markets, parks, and government offices, and it distributes second-hand smart devices to vulnerable groups such as the elderly, low-income families, and the disabled (Agyeman and McLaren 2015, 80–85).

Phase 2 (2015–2019) shifted the emphasis from suppliers to consumers and focused on diversifying the range of actors involved in the sharing activities. Although the efforts of the city improved the spread of sharing practices across Seoul, the policy interventions are still perceived as fragmented by citizens, who are still addressed by the policy as beneficiaries/receivers of the services and not providers. In addition, the lack of non-economic standards in the measuring system of the impact of sharing economy services is a trigger for companies supported through the policy to focus on profit as much as possible.[2]

Phase 3 (2021–ongoing) is designed to restructure the policy in the post-pandemic Seoul. It will also account for the global agenda policy priorities implemented through the Sustainable Development Goals. The third phase of the Sharing Economy Policy is based on ecological sustainability, civic autonomy, and cooperation between city residents to co-manage urban goods and co-produce sharing services.[3] Phase 3 anticipates the city of Seoul investing efforts to promote a shift of city residents from being mere receivers of public support and subsidies to proactive actors who cooperate and partner with the city to co-produce sharing economy services. In this phase, the city intends as well to focus on the ecological impact of the sharing economy.

Other scholars who have studied Seoul's policies have noted that the role of the state is very strong and even "interventionist" in promoting the sharing of urban resources (Agyeman and McLaren 2015). This interventionist role is demonstrated by the fact that the Seoul Metropolitan Government provides direct grants to sharing economy businesses that provide services with a public interest (e.g., books, tools, and children's toy sharing).

Through the last phase of the Sharing city policy, Seoul is working not only to increase the number of businesses in the sharing economy but also to increase social and economic pooling between city inhabitants, with strong state support and coordination. The sharing economy is encouraged and bolstered in underperforming or weak sectors. The city encouraged, for instance, the sharing of rooms as a way to improve access to housing and also to promote intergenerational support, and the sharing of children's clothes. The Sharing city policy is thus an attempt in part to utilize the sharing of urban goods and resources to improve the quality of life for socially and economically vulnerable individuals and communities through the creation of new jobs, the provision of services of public interests, and the strengthening of community ties and solidarity networks. The policy also proved to be instrumental after the COVID-19 outbreak showed the importance of sharing and co-producing services (especially virtually) for the quality of life of urban residents. Most importantly, the city is now working to stimulate civic autonomy, self-organization, and cooperation among urban residents to co-produce sharing economy services that are environmentally friendly and inclusive and contribute to solving urban commons challenges (interview with Seoul City Hall City Transition Division, July 2021).

THE BOLOGNA EXPERIMENT: COLLABORATION PACTS AND NEIGHBORHOOD EXPERIMENTATION

Around the same time that Seoul was implementing its Sharing city policy, the city of Bologna was initiating a policy process to introduce collaboration as a method for governing the city and many of its urban resources. Its goal was also to apply the same design principles animating co-governance of urban resources to other local public policies. As part of what came to be known as the *Co-Bologna process*, the city of Bologna adopted a new regulatory framework on the urban commons. The Regulation for Collaboration between Citizens and the City for the Care and Regeneration of the Urban Commons (hereinafter called the Bologna Regulation or the Regulation) was adopted in 2014 after two years of field experimentation in three city neighborhoods (Iaione 2012, 2015). The process of experimentation that led to the Bologna Regulation was carried out by the city through the Cities as a Commons project. The project was implemented with the support of a not-for-profit foundation, the *Fondazione del Monte di Bologna*, a local NGO, the *Centro Antartide*, and a group of legal and public communication experts from the *Labsus* NGO in which one of the authors of this book, Iaione, served as managing director.

The Bologna Regulation is the first prominent regulatory innovation in Italy, in Europe, and at the international level to introduce collaboration as a method for governing the city and its resources. The city of Bologna has been internationally recognized for this experiment, becoming one of the inaugural winners of the Engaged Cities Award by the Bloomberg Philanthropy's Cities of Service nonprofit group. The award recognizes the most effective city strategies that engage citizens to solve critical problems and that are replicable by other cities around the world.

Bologna is a mid-sized city, the capital of an Italian region with a long history of strong civic engagement and progressive political culture, Emilia-Romagna. As scholars like Robert Putnam have observed, the Emilia-Romagna region is quite exceptional in terms of the presence of civic community and strong, responsive representative institutions (Putnam et al. 1993, 6–7). The city of Bologna has historically been characterized by a high degree of political activism and diversity of civic organizations and dynamism. Especially noteworthy is the enduring presence of social cooperatives throughout the city. The party in power has historically had a strong connection with civic life through institutional mechanisms designed to offer participatory spaces within the government, although this connection started to wane in the late 1970s as these institutional spaces lost some of their dynamism over time (Baiocchi 2003). Even as formal avenues for institutional participation in government had been on the decline, civic life remained robust in the city. In 2010, an analysis of the social economic sector in Bologna counted 1880 civic associations in total, 88 social cooperatives, and 63 foundations (regional data set on foundations) (Eurocities 2011, 7). When the experimental process to design the Bologna Regulation was conceived in 2011, Emilia-Romagna was one of two regions in Italy—the other was Tuscany—with a law promoting citizens' participation, Reg'l. Law 3, 2020 (Regione Emilia-Romagna 2010).

The first step in Bologna's process toward a new regulatory framework for collaboratively governing shared urban assets was a seminar called The City and the Commons hosted by University of Bologna professor Marco Cammelli on December 11, 2011, and attended by local officials and people from both the public and knowledge sectors (Olivotto 2016). At the seminar, Christian Iaione presented a background study that mapped existing examples of Italian city regulations and governance schemes under Italian law, including regulations enabling and supporting community gardens or green spaces; regulations for the promotion of street art; regulations implementing Italian law on microprojects for neighborhood and street-level public space and infrastructure projects;[4] city resolutions for temporary use of city-owned underused buildings; community cooperatives, neighborhood consortia, and participatory foundations incorporated for the promotion of active citizenship, among other legal entities that could be incorporated as an alternative to contractual arrangements. These regulations demonstrated that existing public policies on green spaces, urban culture and creativity, public space place making, and the like were consistent with the idea of the public authority and citizens sharing the responsibility over different kinds of urban resources. For instance, the previously mentioned national law on neighborhood and street-level projects introduced the possibility for local officials to work collaboratively with local civic organizations to realize these projects in exchange for a tax credit and through a streamlined administrative procedure. Although these regulations were not positioned or described as commons-oriented policies, the study showed that the Italian constitutional principles of horizontal subsidiarity and civic collaboration and of legislative actions taken pursuant to those principles provided a rich environment for the implementation of these kinds of policies at the local level (Iaione 2013, 2019).

Following the seminar in 2011, the policymaking process continued with field experimentations in three city neighborhoods as the first step in the process of drafting a new regulation for the urban commons. The goal of these experiments was to generate concrete evidence for and to document the existence of the presence of administrative and legal bottlenecks or obstacles faced by residents and civic organizations that desired to undertake small-scale interventions and street-level projects to improve the social and economic life of city neighborhoods. The results and insights generated by these field experiments created a baseline of knowledge for an internal group of city officials and two scholars, Gregorio Arena and Christian Iaione, to draft the first version of Bologna's regulation on the urban commons. The first draft of the Regulation was then submitted for review by the academic community, chiefly public and administrative law scholars, and a final version was approved by the city council in February 2014.

The urban commons covered by the regulation includes public spaces, urban green spaces, and abandoned buildings and other infrastructure. However, its definition of the commons is quite expansive, directly relating the urban commons concept to the quality of life in the city and the idea of human flourishing:

[T]he goods, tangible, intangible and digital, that citizens and the Administration, also through participative and deliberative procedures, recognize to be functional to the individual and collective wellbeing, activating consequently towards them, pursuant to article 118, par. 4, of the Italian Constitution, to share the responsibility with the Administration of their care or regeneration in order to improve the collective enjoyment.

The central regulatory tool of the regulation is the collaboration pact that establishes the object of care, such as a park or building, and the rules and conditions of collaboration among any group of citizens and the local government, among other actors. The collaboration agreement could be for long-term care of a particular resource or for a single or short-term intervention. City authorities were required to invite proposals for specific pacts of collaboration that were then evaluated and approved (or not) after a co-working phase. The purpose of the review process, according to the regulation, is to ensure that the interventions of city inhabitants for the care of the resource(s) at issue were in harmony with both public and private interests (section 10, par. 2 and 3). The regulation also provides for the transfer of technical and monetary support to the collaboration and offers guidance for defining the borders of the resource to be managed through a collaborative pact. Also important are the stated norms and guidance contained in the regulation, which speak to the importance of sustaining common resources, maintaining the inclusiveness and openness of the resource, protecting the public interest, and directing the use of common resources toward the "differentiated" public. Finally, the regulation speaks of fostering urban creativity chiefly through urban and street art and digital infrastructure. The regulation also anticipates the willingness of inhabitants and property owners to utilize private property for some public uses through "shared management." It specifically supports the creation of street and neighborhood associations, consortiums, cooperatives, or neighborhood foundations collaboratively managed by a "plurality of active citizens" representing at least 66 percent of real estate or commercial activities that are on private space for the public (section 14).

The regulation in and of itself is only one piece of what became known as the Co-Bologna process. Local officials have since enacted other public policies, not based on or directly linked to the regulation, that are part of the policy ecosystem to enable and facilitate city inhabitants and other actors (including social innovators, civic organizations, local entrepreneurs, and knowledge institutions willing to work in the general interest) to enter co-design processes leading to local collaborative governance of an array of urban goods, services, and infrastructure. These other policies include *"Incredibol"* (2011), a public call offering grants for projects of cultural enterprises in the city (funding between twelve and twenty projects per year); a digital platform and civic space, *Comunità Iberbole Platform* (2014), giving city inhabitants access to the collaboration pact proposals and an opportunity to comment and discuss them; a co-design process called *Collaborare è Bologna* (2016); neighborhood laboratories to plan for using abandoned or underutilized public assets to install collaborative spaces; a policy to stimulate economic development at the neighborhood level, *Pilastro* (2016); and *Participatory Budgeting* (2017) in which residents choose how to allocate and spend the public budget.

Another core component of Bologna's move toward being a collaborative city as part of the Co-Bologna process is the establishment of an enabling institution. In 2018, the city decided to change the legal structure and name of the Urban Center of Bologna. For almost twenty years the Urban Center had been managing urban information and communication tasks. It was incorporated as an association. The legal structure was changed into a foundation and therefore recognized as a legal person. The name was changed into Foundation for Urban Innovation. The foundation is essentially controlled by the city of Bologna and the University of Bologna and participated by several other urban key public stakeholders such as the metropolitan city of Bologna and the local public housing agency. The Foundation conducts its activities on the ground through the support of an interdisciplinary team, the Office of Civic Imagination (Fondazione Innovazione Urbana n.d.). The Office for Civic Imagination is structured as a multi-disciplinary team working closely with the municipal administration to find innovative solutions to common urban problems and to implement those solutions in accordance with the principle of civic collaboration. The team supports all citizen-led processes, collects and disseminates public data, aggregates skills and tools, and supports collective practices of city inhabitants including forms of cooperation and resource integration toward the establishment of collaborative economy ventures at the neighborhood level. The Office for Civic Imagination also supports the participatory policies of the city, such as participatory budgeting or the participatory urban planning process. These policies are implemented through neighborhood labs intended by the local administration to provide spaces for "co-design" rather than just the standard "participatory" process that characterizes many municipal processes (Iaione 2016). This work is supported by a team of Agents of Proximity (six agents for each neighborhood). (Ginocchini and Vai 2021; D'Alena 2021).

The Pilastro neighborhood 2016 project is illustrative of how these policies were operationalized in one neighborhood in the city, starting with application of the regulation and then expanding the approach to the larger policy process undertaken by the city, which came to referred to as Collaborare è Bologna. The Pilastro project was designed to stimulate the creation of a "collaborative economy district," rooted in the concept of the urban commons, with the goal of constituting a community–based neighborhood development agency inspired by the model of the French *Regies des quartier* (CNLRQ 2017). Widely implemented in France but with ancient roots in Italy (Laino 2012), the *Regies de quartier* are nonprofit agencies coordinated at the national level but governed at the local level through an association "de loi 1901," the traditional model of a volunteer nonprofit civic association composed of a partnership of urban inhabitants, city authorities, neighborhood level associations, social workers, and other actors relevant to the inclusive economic development of the area. They specifically address local development and community development issues and support residents finding jobs by investing in neighborhood infrastructure. Unlike the *Regies des quartier* model, the Pilastro project includes private and civic actors, a community association, and a community cooperative. The community cooperative would serve as the basis of an institutionalized public-private-commons partnership for neighborhood development (interview with city officer for the Pilastro 2016 process, city of Bologna, August 2018).

In the same vein, the Pilastro neighborhood development agency was created as a collaboration between the city of Bologna, the San Donato Neighborhood administration (the city of Bologna is administratively subdivided in fifteen neighborhoods, whose government is constituted by the Neighborhood President and the Neighborhood Council), the city-regional housing agency (ACER Bologna), a local bank, the local farmers market, the consortium of a local commercial malls, and a not-for-profit foundation.

The city of Bologna assisted and supported the creation of the Association Mastro Pilastro designed to evolve into a community cooperative—and the neighborhood development agency with a specific goal: to stimulate community-based economic development in the North-East District of the city, specifically in the Pilastro neighborhood. The foreign population in the Pilastro neighborhood constitutes approximately 16 percent of the total population, and within that is a very high population of youth (Ginocchini et al. 2013). The neighborhood development agency was designed to help socially and economically vulnerable migrants acquire or improve professional skills, find job opportunities, and integrate them into the social life of the neighborhood and city. It is supposed to accomplish these goals through the pooling of public, private, nonprofit, and informally organized residents' groups and the different assets and resources they bring and devote to the agency's common mission. The Pilastro project after the initial phase of experimentation lost its thrust after a change in the political cycle and it is no longer supported by the city.

Local officials built on these collaborative interventions by designing neighborhood laboratories as places of active participation, care, listening and enhancement of solidarity within and among communities. These laboratories then became a preparatory phase for the citywide participatory budget process. The neighborhood-based experimentation further created a path to establishing the conditions for polycentric governance of urban resources. This is done in part by the creation of new platforms and institutions that give residents more control over the decisions that shape their lives. This experimentalism has led to the creation of an association of commercial shops and artists in one neighborhood (Bolognina) (Co-Bologna n.d.a.), the establishment of a community association in another (Pilastro) (Co-Bologna n.d.b.), and the drafting of an action plan to prepare the work for a participatory neighborhood commission in a third neighborhood (Croce del Biacco) (Co-Bologna n.d.c.).

Another example of an urban experimentation that adopted an approach similar to that of Bologna by using neighborhood labs and experimentation techniques to implement innovations through a citywide laboratory was Mexico City's *Laboratorio para la Ciudad* (Cruz Ruiz, 2021), which was created in 2013 and operated until 2019. The Lab worked as an incubator and co-creation platform for Mexico City inhabitants, encouraging social cohesion and political participation. The Lab provided the opportunity for pooling ideas, improving them through co-creation processes, and then capitalizing them through existing city-funded programs such as the Participatory Budget or the Neighborhood Improvement Program. It also promoted projects aimed at improving the technological capacity of city residents alongside the creation of tech-based tools to improve policies. For example, the *Codigo para la Ciudad* (Code for the City) project sought to develop inexpensive, creative, and technological strategies to implement data-based public policies to improve the quality of life of Mexico City's inhabitants. The program operated through calls for proposals addressed to small enterprises and city residents. *Ciudad Propuesta CDMX* is a digital platform that aims at improving the visualization of ideas and proposals submitted through the city's participatory programs. It serves as a mechanism for pooling and passing along ideas for urban and community revitalization that can be replicated and adapted between neighborhoods and capitalized via the Participatory Budgeting Program or the Neighborhood Improvement Program.

The Bologna experiment certainly represents an important step in the transition of a city in which many forms of urban commons emerge to one in which local officials are enabling and facilitating different degrees of urban co-governance through the creation of nested but autonomous and independent institutions at the neighborhood level. Using Arnstein's famous ladder of citizen participation (1969), the Bologna experiment falls on the high end, representing a level of citizen power and influence unlikely to exist in many other cities and local governments. This measurement, of greater or lesser power and influence by ordinary citizens, could potentially reduce asymmetries in the concentration of political and economic power and thus resist the disruption that results from changes in political cycles, an element that often represents a risk associated with experimental democratic innovations. However, reduction of these asymmetries is not a foregone conclusion, and as ambitious and groundbreaking as the Bologna experiment was, the process reveals various blind spots in trying to instantiate collaboration as a citywide policy and practice.

EVALUATING BOLOGNA AND THE REGULATORY RACE TOWARD THE COMMONS

The Bologna Regulation is emblematic of the constitutive approach to local policies geared toward creating or enabling collective action to govern, manage, or steward shared urban resources. Its success demonstrates that the neighborhood level is where the city can find the most active actors willing and ready to collaboratively share resources toward the creation of collective social and economic well-being. The interventions for the care and regeneration of shared urban resources via the collaboration pacts have spread far and wide across the city. To date (as of March 18, 2022), the Bologna Regulation has resulted in more than six hundred pacts of collaboration signed and implemented since 2014.

As part of its policy experiment, local officials agreed to conduct an evaluation of the results of its policy and specifically to measure the impact of the Bologna Regulation after two years of implementation. It hoped to learn from the experiment, to integrate the lessons learned, and to amend the policy accordingly. So, our researchers undertook an analysis of the collaboration pacts entered as a result of the Bologna Regulation to understand the impact of this policy on urban democratic qualities including participation, deliberation, responsiveness, and accountability (Morlino 2011;

Dryzek 2009). The evaluation process consisted of a quali-quantitative analysis of 280 collaboration pacts approved by the city between 2014 and 2016 (De Nictolis and Pais 2018). Data was collected through analyzing policy documents,[5] surveying civic signatories of the pacts, and carrying out focus groups and interviews. This initial evaluation found that in most cases the pacts were not facilitating deep civic participation, in the sense of drawing in people who have not otherwise been civically engaged or in drawing people together across social differences (De Nictolis and Pais 2018). On one hand, the pacts do reflect the *potential* to activate those not previously engaged at the local level, as in some cases, the pacts were the first opportunity for collaboration among pact signatories who were previously unknown to each other.

The large number of pacts of collaboration signed is a significant and positive outcome of the Regulation, but they lack a diversity or multiplicity of actors. Some pacts are signed by not-for-profit foundations, social enterprises, start-ups, and businesses. Most pact signatories consist of NGOs or informal groups and are bilateral partnerships such as city-NGOs, city-private actor, city-individual resident, or city-informal group of residents. In addition, although the regulation embraces an expansive definition and scope of application of the urban commons—that is, opening up the possibility of collective governance and areas of intervention related to social innovation, urban culture and creativity, collaborative services, and digital innovation—the majority of collaboration pacts are focused on activities of resource care, such as the removal of graffiti from a wall or the cleanup of a public street or park, or the co-management of spaces or buildings to realize cultural and social activities such as artistic exhibitions, workshops, and laboratories to transfer creative, digital skills to neighborhood residents. Examples are street art days organized in a public space (*Street Art Pact at the Zaccarelli Center*). Some of those pacts involve digital tools such as *OUTakes archive*, a pact to collect and digitize material linked to the LGBTQ movement's history.

There are, however, a handful of pacts dedicated to complex urban regeneration projects. The *Bella Fuori* 3 pact, for instance, involves the renovation of a square in a low-income neighborhood coupled with the creation of an NGO and residents' associations for the co-management of the green areas and the playground built on the square or that involve city-owned buildings to create collaborative economies or urban welfare services. Another example is the *HUB Underground Base* pact, involving a large group of NGOs dedicated to renovating a building for cultural and creative activities, learning labs, workshops, and incubation of neighborhood projects that support networks of residents working together. The pacts signed, for instance, with the organization Social Streets involve resident networks composed of neighbors who live on the same block and that help coordinate maintenance of their streets. The pact *Shared Management of the Ex Serre Giardini Margherita* involved the cooperative Kilowatt's creation of a community garden inside a neighborhood park that was designed to include an affordable co-working space. In the pact *Forever Ultras*, an NGO agreed to use its own resources to renovate and manage a city-owned building that offers socialization opportunities for elderly people through sports or artistic exhibitions as well as an archive of football-related materials. Another example is the pact *Piantala*, which created a circuit of circular economy in which an NGO collects unsold plants from hatcheries and distributes them to city residents who are willing to turn a destitute green area into a community garden (Iaione and De Nictolis 2021). The institutional platform of the Bologna Regulation offers these and many more examples of pacts of collaboration at work (see http://partecipa.comune.bologna.it/beni-comuni).

The results of the analysis of the first phase of implementation of the Bologna Regulation call into question whether the purpose of most of the collaboration pacts was to facilitate the creation of nested but autonomous and independent institutions or collectives that allow residents to steward shared resources. The results also seem to validate the skepticism that collaboration pacts can reduce asymmetries in the concentration of civic power and influence, particularly given the social and economic inequality and geographic stratification that characterizes urban environments. As others have observed critically, the implementation of the Bologna Regulation has allowed the city to benefit from engagement with the social sector and active city inhabitants to maintain public spaces and infrastructure, which were visibly in decline as a result of the economic crisis and the privatization of public resources (Bianchi 2018). At the same time, without a stronger political valence, the invitation for local inhabitants to take care of and regenerate urban resources sidesteps the hard work of deep civic collaboration and the need to transition to new economies that include and can empower many residents living on the social and economic periphery of many cities (Bianchi 2018, 301–302).

Of course, further evaluation is necessary to determine if these initial results from the first 280 signed pacts are also observed in the additional pacts signed since 2018. We have not evaluated nor followed the results of the experiment after conducting the initial evaluation. Since our involvement, we anticipate many improvements to the Bologna experiment arising from a combination of several factors: evaluations and reporting conducted by the city administration[6]; further investments by the city government and the administration on the improvement of this policy; the efforts produced by proactive and motivated residents of Bologna; the activity of scholars and advisers involved in the Bologna process. All of these are likely to have resulted in improvements in implementation of the policy, as well as remediation of the weaknesses that we identified after the first, experimental phase of a pioneering journey.

THE REGULATORY RACE TO THE COMMONS IN ITALY AND OTHER EU CITIES

Given the evaluation results, and on the heels of the Bologna experiment(s), one troubling development has been the adoption of the regulation by other cities, creating the danger of reflexive replication of the law in the absence of learning the lessons from Bologna's experiment. Several cities in Italy adopted the Bologna Regulation, engaging in a sort of regulatory race toward the commons without much consideration of how creating a city based on commons principles might differ (even in the same country). This uncritical adoption of the regulation alone, without the other aspects of the Co-Bologna process, has given rise to the understandable critique that the Bologna Regulation entails a top-down, paternalistic approach to the commons even if it arguably represents an enlightened from of civic activism (Mattei and Quarta 2015).

Notably, other Italian cities have chosen a different path, avoiding this regulatory race and engaging in more context-specific approaches to the city as a commons, as we discuss subsequently in this chapter. Examples include Naples, which forged a more declaratory approach in contrast to Bologna's constitutive approach, and Turin's own interpretation of Bologna's constitutive approach.

A sort of regulatory race was triggered by the Bologna experiment also at the EU level as cities like Amsterdam, Ghent, and Madrid demonstrated interest in the Bologna experiment, likely to re-interpret or update their long-standing tradition in participatory approaches or third-sector stimulus policies. They each invested to different degrees and with varying results in adapting the Bologna approach to the local context in order to craft their own unique piece of urban law for the urban commons.

We briefly mention here two examples of these non-Italian city approaches inspired by Bologna's regulation. Both the examples come from EU cities. The first is from Ghent, Belgium's second largest municipality and a city known for, among other things, its participatory approach to civic life. Consistent with this participatory orientation, Ghent began the process of transitioning to a *commons plan* in the city after it invested time and resources to study existing regulatory and nonregulatory approaches to stimulating urban commons, particularly in European cities, and their adaptation through experimentation to the local context. Ghent also engaged in a mapping exercise to assess its existing *commons projects* in the city, which numbered over five hundred, and also conducted interviews and held workshops with those already involved in the creation of urban commons. The second example comes from Madrid, which by contrast began its process by passing an ordinance for public-social cooperation (hereinafter called the Madrid Ordinance) that very closely mimicked the Bologna regulation and more generally the urban commons regulatory movement that was flourishing at the time (Ayuntamiento de Madrid 2018a). Most notable in the Madrid regulation is the Preamble, which recognizes civic collectives as part of an obvious nod to using the framing of the urban commons momentum to reinforce and give a new legal language to the long-standing tradition of public-social partnerships between the city and civil society organizations, that is, the so-called third sector.

The Madrid Ordinance distinguished between the *co-management* of social activities and the *social management* of urban assets. The Ordinance allowed the city to grant the use of publicly owned buildings and spaces to nonprofit entities to carry out activities of public and social interest that produce a positive impact on the community and that foster stability and continuity (article 4). A public call or tender procedure is used to select appropriate co-creation and co-management projects and the entities involved, which are given the exclusive right to use the building or spaces. The city administration at the time claimed that the Ordinance was a pathway for the involvement of individual actions and citizen cooperation to change the development of the city itself. Collectives' efforts would be geared towards common interests defined by a multiplicity of actors coming together in a cooperative approach to city making.

The public-social cooperation in the Madrid ordinance can be implemented through legal tools already existing in the Spanish legal landscape (e.g., the authorization or grant of use of goods and buildings to nonprofit entities for social purposes pursuant to law 33/2003 of the Public Administration Assets Management and the Regulation on Local Entities Assets Management approved through the Royal decree 1376/1986). The key takeaway from the Madrid Ordinance is that it was an attempt to nudge social collectives to move beyond mere participation and assembly decision making in local politics to a more collaborative or cooperative mode that would bridge local innovation practices with the long-standing tradition of democratic cooperation. The hope was that by partnering with the city administration, local collectives and/or social cooperatives would develop projects of common interests. The idea was to adapt and expand old methods of social cooperation to new forms of urban collaboration. The legal framework would reinforce both the old and the new forms of cooperation through claiming recognition for urban commons (Sobral 2018).

After the elections in May 2019 the new governing coalition declared their intention to either profoundly modify or ultimately repeal the Ordinance, calling into question its ultimate impact (*Europapress* 2019). However, limited in terms of legal innovation and city inhabitants' engagement or self-empowerment strategies, in the end Madrid did take the bold initiative to recognize, in passing the Ordinance, the right of collectives to be part of the public-social cooperation legacy of the city. Nevertheless, the constitutive approach exemplified by the Madrid Ordinance does not leave much room for legal experimentation and ensures that city officials continue to hold the reins of civic engagement processes. In 2020, the city government launched a consultation on the ordinance before repealing it with the intention to partially integrate it within a Framework Regulation on Civic Participation (Ayuntamiento de Madrid 2020a). The goal of this change would be to integrate, in a single regulatory text, the different mechanisms of participation, for the sake of a better, more effective and efficient system of city inhabitants' participation in the management of city-owned assets (Ayuntamiento de Madrid 2020a). The city government agreed to hold the public consultation which took place on the Decidim platform between January 30 and February 3, 2020 (Junta de Gobierno de la Ciudad de Madrid 2020). The consultation process received only 124 comments. Almost all modification proposals received positive answers (Ayuntamiento de Madrid 2020b). However, to date, no further initiative to pass the framework regulation has been taken. It is not clear whether the mechanism of public-social cooperation as established by the ordinance is still in force.

In contrast to Madrid, Ghent's constitutive approach began by innovating its already deeply participatory and collaborative regulatory landscape. Ghent's political appointees and administrative staff are steeped in a culture with a long tradition of participatory and co-creation approaches. In 1998, for example, the administration created a unit that enables policymakers to integrate a bottom-up approach to planning and decision-making processes. The unit still exists and has developed different instruments (e.g., participation platform, crowdfunding, temporary use, and participatory budgeting) to work closely with city residents to support their neighborhood initiatives. The city has also been experimenting with the temporary use of brownfield (contaminated former industrial) sites and empty buildings for over a decade in the development of urban renewal projects. The city therefore has a history of allowing the temporary use of the sites and buildings by the residents. DE SITE, the first temporary use ten years ago, included allotments, a greenhouse, two urban horticultural plots, a football field, a bike playground, and an urban farmstead created on the site of the former Alcatel Bell factory in the Rabot district. The city of Ghent partnered with Samenlevingsopbouw Gent (*Community Development Gent*, a community development foundation working on social housing, social work, and social protection projects; see https://samenlevingsop bouwgent.be/) and started the project to get residents involved in their district and to stimulate a public discussion on the forthcoming urban renewal process. The city provides subsidies to initiators of temporary-use projects via a temporary-use fund. Every year, the city council allocates a budget of €300,000 for this purpose (URBACT 2018).

A natural evolution of the city of Ghent's approach to engaging citizens in urban renewal was its decision in 2017 to develop official guidelines for the development of commons governance throughout the city. Its Commons Transition Plan (strongly supported by the city's mayor and political coalition of leading parties at the time), cowritten by a group of two hundred local activists, pushed to expand and facilitate bottom-up commons initiatives (Bauwens and Onzia 2017). The stated purpose of the city's approach was to undertake a broad study of the landscape of commons-oriented projects in the city and to understand the opportunities inherent in active city inhabitants engaged in co-constructing new initiatives and projects in response to urban challenges. The process resulted in recommendation of a plan for the city to continue to facilitate and strengthen citizen-led initiatives and to make the city a partner in those initiatives. The Commons Plan recommended a cross-sector institutional framework for supporting these partnerships, which included a city laboratory that would prepare a Commons Accord between the city and the citizens' initiatives, to be modeled after the Bologna Regulation; another partnership was the establishment by the city of multi-actor alliances.

The aim of Ghent's involvement in the EU-funded Civic eState network would be to craft a regulation specifically designed to promote Ghent's urban commons, starting from the guidelines provided in the Commons Transition Plan and building on that using the existing tools that had been adopted in the previous years by the city to promote the projects of community-based and self-organized use of spaces mentioned previously.

The city is currently working to initiate two pilot projects as part of its larger plan. The first is Saint Joseph Church, a desanctified church located in the Rabot-Blaisantvest neighborhood (one of the poorer neighborhoods of the city) purchased by the city in 2019. The city is willing to generate a public-civic management of the building using several tools (including an open call drafted by an interdepartmental working group to find a project manager and a real estate agreement signed by the manager and the city). The second pilot project is a tax exemption for the use of public space for citizen and neighborhood initiatives. This was inspired by the Bologna Regulation, in the part that provides the possibility for cities to allow a tax exemption for the signatories of the pacts of collaboration.

Although the city does not have a uniform regulatory framework to support/regulate city residents initiatives and public-civic collaborations/partnerships, it is actively participating in the network with the goal of crafting an urban policy inspired by the design principles adopted by Naples for civic uses. However, the city is still in the learning and experimentation phase. Ghent intends to start from a few legal instruments already in existence to support/regulate citizen initiatives and public-civic collaborations/partnerships, such as agreements (e.g., subsidy agreements and real estate agreements), city regulations (e.g., subsidy regulations), and permits for the use of public spaces.

Real estate agreements are entered into by the city of Ghent (real estate department) or by the autonomous municipal company of urban development of the city of Ghent and third parties concerning the transfer of ownership or the right of use of real estate. Different agreements are anticipated by the city of Ghent, such as rental agreements, management agreements, occupancy agreements, lease agreements, and agreements for temporary use. The agreements stipulate the term/duration of the agreement; compensation/fee for the use of the building; costs of utilities: water, gas, and electricity; maintenance costs; and insurance (e.g., fire insurance) and guarantees. In the case of citizen initiatives of public-civic collaborations/partnerships, the choice of contracting party for these real estate agreements is generally the result of an open call.

Another tool to promote citizens' initiative toward the care of the urban commons are city subsidies. The conditions under which citizens and organizations can be entitled to subsidies from the fund of temporary use are the following: approval of the owner of the site; the project's contribution to an increased quality of life in the neighborhood; importantly, the self-initiative, co-management, involvement, and creativity of the applicants; and a commitment to realizing the project within two years from the approval. Subsidy agreements and real estate agreements are used by the city of Ghent to support and regulate citizens' initiatives and public-civic collaborations/partnerships.

The experience of Ghent showed that a process to promote co-governance of city-owned buildings must be accompanied by open procedures to advertise the opportunities as much as possible. These procedures should be formalized and promote competition while also being as flexible as possible in order to enable the emergence of urban innovations and stimulate new actors to participate. An example is the Open Call published in March 2021 by Ghent to identify a coordinator to become involved in the management of the Saint Josepf Church. The church was purchased by the city after a previous URBACT project on temporary use and the coordinator will be responsible with drafting a management plan of the site by city residents (Civic eState 2021b, 54; URBACT 2018).

THE DECLARATORY APPROACH: URBAN CIVIC USES AND CIVIC MANAGEMENT

Other local governments are pivoting from the constitutive approach represented by the Bologna regulation and those it inspired, toward more carefully designed policies adapted to unique local contexts. This development appears to represent a turn away from the regulatory race toward one model of supporting the development of urban commons throughout a city. The examples of Naples, Barcelona, and Amsterdam discussed in this section are expressions of the second approach to city policies: the declaratory approach, which recognizes the emergence of urban communities willing to self-organize in order to utilize and manage city assets and infrastructure to produce common goods and services, instead of providing them first with a regulatory or policy framework to do so. Like the constitutive approach, these policies situate the local government as an enabler of collective governance while also facilitating public-community partnerships and the development of community-based enterprises.

The declaratory approach recognizes the community right to self-organize to assert control or management of key assets for local communities. For instance, a few Italian cities have revisited the ancient legal category of *civic uses*, recently revived by Italian scholars to identify a property right beyond strictly public and private categories (Grossi 2017; Marella 2012). The concept of civic uses embraces the right of communities or collectivities to use and collectively manage land or structures under the control of local authorities. Local regulations adopting this category allow those initiatives to flourish as self-organized, new urban collective enterprises constructed from underutilized and abandoned city assets.

An important precursor to these contemporary policies can be traced back to the English community right to bid (hereinafter called CRTB) and community right to challenge (hereinafter called CRTC) laws, which are a part of the UK Localism Act of 2011. These policies enable local communities to proactively self-organize, to claim a local asset that produces social value, or to manage a local public service to suit their needs. Part 5, chapter 3 of the Localism Act along with the Regulations 2012 (SI/2012/2) (hereinafter called the ACV Regulations) introduces the Assets of Community Value (hereinafter called ACV) and deliver a *Community right to bid*. They provide the possibility for *local communities* (organized in various ways, such as a parish council, a community council, an NGO, or another form of voluntary association) to require listing as an ACV of an urban asset, which can be a building or a portion of land, usually a neighborhood commercial activity such as a local pub or a village shop; a theatre or a cinema; a health care center; or a community center. The request is made to the city council to prevent the closure or sale of an urban asset that produces value for the local community. Once the asset is listed as an ACV and the owner decides to sell it, the owner must notify the public authority and allow a community to submit a written request to be considered as a bidder. The community can benefit from various forms of support provided by the city to raise the necessary funds. The support can come in the form of administrative support (such as advice on how to raise the purchase price or how to conduct negotiations) or technical support (such as advice on how to hire and supervise staff, how to maintain the building, or how to take a loan). Support can come even in the form of small grants by the city (Samuels 2017, 485).

The CRTC, similarly, is a way to re-envision the delivery of public services and open the door to the proactive role of local communities. The CRTC empowers voluntary and community groups, parish councils, and employees of local authorities to express an interest in taking over a public service currently delivered by the local authority. The low utilization of the CRTC across communities in England suggests that when an advanced declaratory approach is not coupled with strong capacity building and support from the state, and without innovating legal procedures and tools that allow cities and communities to work together, there is a risk of private sector exploitation. The bureaucratic complexity and costs associated with the process (that eventually ends up in a public procurement process) carries the risk of private stakeholders using the process to achieve the goal of privatizing the service (House of Commons 2015, 22–23; Layard 2012, 141).

Appreciating the UK experience may help to understand why some cities have created legal avenues for communities or coalitions of local actors to claim a sort of pre-emption right and pre-existing privilege over some assets that were once critical and essential to the livelihood or vibrancy of a city, neighborhood, or village and at some point become neglected and abandoned. The UK provisions are an important precursor to the policy from Naples that we describe, along with other cities that have adopted a declaratory approach. These policies instantiate legal recognition of a pre-existing right of communities to act in the general interest. The UK policies have paved the way for cities like Naples, Barcelona, Amsterdam, and other cities that desire to make a bold legal statement by introducing strong community rights. There are two important distinctions between the UK policy and the policies that we describe here. The first is the role of the public sector, and the second is that in most of these cases the regulatory initiative comes from the city and not the national level.

The city of Naples has become exemplary of a local regulatory approach that recognizes and embraces the right to collective uses of urban resources and to community self-governance through the creation of cultural commons (Dardot and Laval 2015; De Angelis 2017). The Naples expression of the movement on the urban commons (which stemmed from the national movement on the commons in Italy but took a different path) and then the urban policies that the city issued to support them show how the Naples experience is aimed at addressing the socioeconomic vulnerability of the population (UCLG 2018). Naples is the third largest city in Italy with a population of over 3 million and is also the main city of southern Italy (ISTAT 2020). It has an unemployment rate of 24 percent (the national average is 9 percent), and the average income of its inhabitants is among the lowest in Italy (ISTAT 2021).

The city of Naples emerged as one of the central stages of the wave of public debate and dissent against the potential privatization of water management in Italy, an effort widely opposed by a network of social movements and activists (Mattei and Bailey 2013), and more specifically the decentralization and delivery of local public services or welfare (Lucarelli 2011). The movement for water as a commons in Italy (which emerged in the early 2000s and culminated in the national referendum to establish public management of water in 2011) highlighted the need to establish a public management of water services. In Naples, the output of the referendum was pushed even further and the city implemented changes to the local delivery of water, establishing rules in the bylaws of the city-owned water management company (Acqua Bene Comune [ABC] [water as a commons]) to allow for citizens' close control over the operation of the company (Mattei 2013; Mattei and Quarta 2014; Lucarelli, 2017).

The issue of citizens' involvement in the management of water services collided with another wave of cultural workers (Cirillo 2014) and social movements' action in Naples, around the issue of cultural and creative spaces. Starting in 2011 with the election of former judge Luigi de Magistris as city mayor, the city boosted collective action around cultural commons in the city. First, the mayor appointed the first Italian deputy mayor for the commons, created the legal category of the commons in the city bylaws, added public participation rules into the governance of the water utility company, and then adopted a series of local resolutions that recognize "urban civic and collective uses" of public buildings. We focus on the latter since it is of major interest for the purpose of this book, although we wish to stress that policies for urban commons might work better within a broader policy framework that encourages collective action and public-civic collaboration.

Most of the urban commons that the city has recognized with its policies are aimed at providing housing and urban welfare services (i.e., legal counseling and health services for migrants or low-income people) or cultural activities, often with the goal of raising popular awareness on issues such as migrants' inclusion, gender equality, cultural diversity, and social welfare services (De Tullio 2018; De Tullio and Cirillo 2021).

This process is the result of a transformation of various social movements that have built a "space of alternatives," characterized by the objective of creating opportunities and the satisfaction of needs addressed to a wider sense of community, not only claiming new rights but trying to realize them concretely through direct actions and a community welfare system (Micciarelli and D'Andrea 2020). The process of bringing the Naples regulation into being began in March 2012 in a clash, following a dialogue, between the local administration and activists at the Ex-Asilo Filangieri, a huge former convent occupied by cultural and artistic workers, which resulted in the drafting of a resolution recognizing the right to collectively use and manage the building. This resolution eventually was adopted by the city council, and later resolutions in 2015, 2016, and 2021 recognized seven other public properties as "emerging commons, perceived by city residents as civic flourishing environments and, as such, considered by the city as assets of strategic relevance."[7]

One key component of developing the Naples resolutions was the establishment in 2017 of a renewed Observatory of Urban Commons and Participatory Democracy designed to be a platform for deliberations and negotiations around existing collective use and management of occupied buildings in the city. The Observatory provided a site for local officials to collaborate with communities or users who were informally managing occupied buildings and to participate in co-working sessions to design the resolutions for the recognition of civic uses. These laboratories are an important feature of both the declaratory and constitutive approaches to enabling and supporting urban commons throughout a city. However, the purpose and features of the labs operate somewhat differently in the distinct approaches. In the declaratory approach, as in Naples, these labs are an end point versus a beginning point in the collective or collaborative process of governing a local resource or service. In other words, the labs do not play a role in facilitating or self-organizing activities as they do in Bologna and in other examples. Rather, self-organization occurs through various bottom-up processes by participants themselves who later join with city officials and other actors to collaborate and pool resources.

In Naples, for example, the process of creating the collective uses of local resources arises more out of the intensive use of assemblies, fora, and other participatory decision-making mechanisms. This approach was injected into the Urban Civic Use Regulation crafted by the Asilo Filangieri urban commons. The resource users, in close cooperation with city officials, designed an institutional model revolving around three main organisms: the *management assembly*, the *steering assembly*, and the *thematic tables*. The thematic tables (for example, theater, the visual arts, filmmaking, sound media, self-government, a library, or a community garden) operate as spaces for deliberative discussions on thematic issues. Those interested in proposing or performing activities using the spaces, human resources, and infrastructure of the Asilo can submit a proposal to the management assembly of the thematic table. The discussion takes place during public meetings in which methodologies of consensus decision making, practices of care, and other democratic techniques are adopted (Federici 2018; Micciarelli 2021). Similarly, a steering assembly exists to define the general guidelines or rules for the chosen activities, to approve fundraising and crowdfunding initiatives, and to oversee expenditures and other economic management decisions,

The Naples resolutions embody a declaratory approach by recognizing the right to civic use of abandoned and underutilized buildings and land owned or controlled by the city to communities that are already managing them informally. The mechanism for official recognition is the agreement between the local administration and communities, or collectivities, managing underutilized or vacant land and space. This agreement, termed a Declaration of Civic and Collective Use, lays out the norms for use, accessibility, and governance of the spaces. These spaces are occupied and regenerated by informal communities, contributing to their regeneration largely through self-funding.

With a methodology called *creative use of law* or *legal hacking*, "instead of attributing a concession to an association, the Neapolitan municipality has recognised this structure as an emerging commons, considering first the non-exclusive right of the inhabitants to use them. The assembly ecosystem therefore performs the dual function of organisational-relational model and is recognised by the municipality as the management body. In this case, both collective land governance systems and commons remind us that the priority is not to identify one or more juridical subjects that hold governance powers, but the governance system itself" (Micciarelli, 2022).

The Declarations require that the use and regeneration of these buildings must be directed toward "civic profitability," and therefore it should not be driven by economic or aesthetic ends. These civic assets can be conceived as part of the civic patrimony of the city of Naples, albeit co-utilized and co-managed by city inhabitants, toward the realization of activities pursuing the *general interest*. The public-owned and city-owned assets thus play a central role in Naples's fostering of new forms of social economy through collective planning (Masella 2018) and civic uses of public assets, giving strength to new forms of social inclusion (Turolla 2020). The city is currently working, also through the support of EU funded programs such as the Naples-led Civic eState URBACT transfer network, to encourage forms of economic sustainability that can ensure the long-term survival of the urban commons through, for example, innovative financing schemes (e.g., microcredit, social outcome contracting and social impact investments) Therefore, the valorization of the municipal assets can be understood as a process by which it is possible to confer a greater social and economic value to public assets through their collective use (see www.commonsnapoli.org).

The most recent initiatives taken by the city are intended to tackle some of the weaknesses that normally affect the long-endurability of the urban commons such as the financial sustainability of managed spaces, the identification of means to generate revenues for their maintenance, and the economic sustainability of social and cultural initiatives run in these spaces. This resolution, in particular, encourages city inhabitants to design and submit pilot projects for the improvement of underused and disused municipal assets that can be redeveloped and transformed to new social uses such as health care facilities; reception centers for migrants and asylum seekers; educational gardens, urban gardens and farms; playgrounds for children and teens; artistic installations/exhibitions; activities aimed at promoting "urban creativity"; and regeneration of public spaces as "civic flourishing environments." By enhancing civic actors' role, the local administration wants to promote new forms of *urban civic communities* and to define innovative schemes of public-community cooperation to gain the interest of long-term investors. In this way, the designed civic development environment would become a driver to boost the overall economic sustainability of the process and to promote innovative financing schemes (Iaione 2019).

The city of Barcelona is another example of a declaratory approach; it is similar to Naples but with a more expansive reach across policy silos such as mobility and housing. Barcelona recently enacted a policy on civic heritage and civic uses but with a stronger emphasis on economic and environmental sustainability of collectively managed city-owned buildings (Barcelona City Council 2019a, 2019b). The policy is a declaratory approach that evolved over time, building on a tradition of several years of pilot projects to grant local NGOs the use of Barcelona city-owned buildings to carry out social activities. The institutional orientation and capacity of the city of Barcelona toward public-social collaboration was already high because of the tradition of the *Barcelona model* and the city's long history of collaboration between public, private, and community actors in city governance (Blanco 2009, 2015), as well as the support for co-managed city-owned spaces and infrastructure by nonprofit organizations (NGOs) (Castro et al. 2016). However, these co-managed spaces, although accompanied by their transfer to NGOs, have not been supported by a single policy framework.

Beginning in the 1980s, Barcelona embraced civic management of city-owned facilities and services (Blanco 2009) through individual contractual arrangements or agreements in the absence of a legal framework or of a uniform commitment throughout the local administration. The collaboration between local government and social actors was instead largely managed through ad hoc actions or disparate policies. These actions and policies ended up laying the groundwork for a more comprehensive approach to collective management of city assets—the civic heritage policy that grants the recognition to and facilitates NGOs and city inhabitants to co-manage city-owned buildings for the purpose of producing and delivering social and cultural services (Castro et al. 2016). This policy is just one of a wide range of policies promoting a commons-oriented approach to city governance. The other policies address issues such as affordable housing, pedestrianization of streets, increased public space, energy sovereignty, the social and solidarity economy, digital democracy and justice, and socially responsible public procurement (Blanco et al. 2020). In each of these areas, the current administration is applying the principle of public-civic collaboration embraced by the civic heritage and civic management policies through a declaratory approach of collaboration.

In 2017, the city decided to create a program called Citizen Heritage of Community Use and Management (*Programa de Patrimoni Ciutadà*) aimed at creating a conceptual and normative framework for the promotion and development of the community management of underutilized buildings. The civic heritage program created a special entity, the Citizen Heritage Board (*Taula de Patrimonio Ciudadano*), to oversee and centralize the process of community use and management of municipal assets. An example of the program is its application to the establishment of *civic centers*. These are city-owned buildings distributed throughout the city, the management of which is operationalized in one of three different ways: (1) co-management through shared responsibility and duties between the local administration and a particular NGO; (2) civic management whereby the building is managed by residents but the services offered are provided by the city; or (3) community management whereby an NGO both serves as the building manager and offers direct services. The city also pays for the water and electricity expenses and sustains all expenses necessary to secure the safety of the space. In 2015, there were more than thirty-five spaces governed through civic management including civic centers, neighborhood houses, community centers, cultural and sport facilities, and historic, architectural, and cultural heritage places (Castro 2016, 46).

The civic heritage program was later codified in the "Strategic Plan for Citizen Assets | 2019–2023 New agenda for policies that foster public-community collaboration." According to this plan, the Citizen Assets program consists of three elements: (1) the property transfer for community use and management of buildings or plots of land, realized through a transfer agreement or a transfer contract; (2) the community management of local facilities (cultural, social, sport, and environmental facilities) that are transferred to a community through a collaboration agreement with a multiyear subsidy and in which the facility becomes a city service managed by the community; and (3) community management of citizen-initiative, public-interest services. In the last case, the city provides support for projects initiated by the community that can be considered public-interest services. The city and the community sign a collaboration agreement, by which forms of economic support, technical guidance, and service incubation and acceleration can be provided.

The legal tools that Barcelona uses to collaborate with local groups or communities in the buildings' co-management derive from different legal sources: assignment and permission to use pubic assets for not-for-profit entities (in the Municipal Charter of the City of Barcelona, the Local Entities Assets Regulations, and the Municipal Body Regulations), the collaboration agreement for the civic management of municipal services by nonprofit entities (in the Municipal Charter of the City of Barcelona, the Local Entities Assets Regulations, the Citizen Participation Regulations), and the Procedure Acts for assigning public assets for use and in civic management collaboration agreements (disciplined in the Municipal Charter of the City of Barcelona, the Local System of Catalonia and the Law on Public Sector Contracts) (Barcelona City Council 2017b; Civic eState 2020). One of the lessons of the pilot projects in the first phase of implementation has been the necessity to craft an appropriate legal-administrative framework to discipline the public-community collaboration through a mixed management and co-management model, which could result from the legal interpretation of the public procurement for indirect management of services as well as the citizen participation regulations. One focus of the model would be the economic sustainability of community management.

A key aspect of the Citizen Assets program is community balance, a self-assessment tool provided by the city and drafted by the community that is aimed at quantifying the impacts (e.g., social and environmental) of their activity and the economic value produced by community management. It analyzes matters such as ties to the territory; social impact and return; democratic, transparent, and participation-based internal management; environmental and economic sustainability; and the care of people and processes. When organizations register with the Citizen Assets program, they agree to be included in the Community Balance Sheet. This is carried out through a form containing questions and indicators aimed at assessing joint social responsibility for the project, local identity, democratic management, and focus on the needs of the community and the environment. The Community Balance Sheet: (1) evaluates whether the project is oriented to the needs of the territory (neighborhood) and/or whether it has ties to local networks, other social projects, and territorial platforms; (2) foresees indicators that assess the social impact and the expected positive externalities and reveal the presence of external beneficiaries to the project; (3) focuses on indicators that assess the grade of internal democracy of the governance structure; and (4) evaluates the gender equality, the incorporation of gender perspective, the environmental sustainability, the labor quality, and the promotion of diversity. The program includes a Citizen Assets Board, an internal municipal body whose mission is to ensure the coordination of any transfers of use carried out and to organize a citizen property promotion policy (Mercat Social 2020). An example of the use of this tool is one of the first regeneration pilot projects under the Citizen Assets program. The city granted the use of a building to the Can Batlló Self-Managed Community and Neighborhood Space Association through a "concession of private use of a public asset," illustrating several similarities to the Naples approach with the Declaration of Civic and Collective Urban Uses but with a stronger emphasis on economic sustainability of the collective management scheme (Barcelona City Council 2017a).

The Barcelona City Council Heritage Services drew up a Schedule of Clauses that ensures openness and participatory use of and collective management of the resource, similar to what the Naples Declaration does. The clauses specify that the concession holder shall pay an affordable rent of €650 annually (article 4). The Community must draw the Community Balance every two years and report its results to the city council, as well as all the information requested by the city council regarding the project's governance and activities. Another similarity to the Naples case is that the city council will pay utility costs up to previously agreed limits, on the basis of economic and environmental sustainability. Both parties undertake to ensure that Civic Heritage criteria are complied with and that the Can Batlló concession space is a space that is open to any city resident or organization (article 7) (Directorate of Heritage Legal Services 2018). The city of Barcelona also included the Can Batlló experience of bottom-up appropriation of the space in the urban planning process (Rossini and Bianchi 2019), introducing qualitative changes in the plan of the area, specifically giving a public area designation to new facilities, and improving the design of public spaces (Rossini and Bianchi 2019).

Amsterdam is another example of the declaratory approach that leverages technology to ensure democratic and inclusive city governance. Amsterdam has pioneered a model of an inclusive smart city in Europe through the Amsterdam Smart City program. The city operated in the absence of a national normative framework on smart cities, although governmental actors were constantly involved. The creation of the network of public, private, and social actors necessary for a smart city was fueled by EU funds, but the strategy then proceeded without them (Raven et al. 2017). The city of Amsterdam joined the URBACT Civic eState network with the goal of innovating its smart city strategy, focusing on the promotion of digital social innovation and developing innovative ways of working together with citizens to create public value through a co-city framework. The starting point was the Room for Initiative program, which began in 2010 (City of Amsterdam 2019).

The program calls for residents to submit proposals on six systemic challenges as a way of changing the functions of the local administrative bureaucracy:

1. *Integrated financing*: Making the subsidization of cross-sectoral initiatives structural, in order to quickly respond to multifaceted initiatives (e.g., youth, culture, and public green spaces).
2. *Increasing the sustainability of informal care*: Strengthening the sustainable provision of informal care through professionalization, financing, and accountability in cooperation with the existing care providers.
3. *Right to challenge experiments*: Residents can challenge the city to take over (part of) a regular government task, including resources and responsibilities. The city can also challenge initiators to take over a regular task.
4. *Entrepreneurship in/out of social benefits*: Beneficiaries of social benefits and refugees can now receive a generous allowance for their volunteer work, and travel expenses are reimbursed.
5. *Real estate*: The goal is to make real estate in the city more accessible for social initiatives.
6. *Livability*: The Development Neighborhoods plans in the North, Southeast and New West link to what is already available in the neighborhood and involve city residents in the initiatives provided by the programs.

The legal tools that the city is currently using to collaborate with communities are not defined in a specific policy for the commons but can be retrieved from different parts of the legal framework, and the Ordinance will update them. Examples are zoning plans that will be replaced by environmental planning and might foresee the assignment of open spaces and mixed functions in areas where urban commons can grow, and a vacancy ordinance that offers the chance to prioritize the creation of an urban commons instead of allocating the use of vacant buildings or vacant urban public spaces in the city. The city is currently working on an ordinance specifically for the urban commons that, in combination with the social value measurement system and the public-community financing scheme, would provide an appropriate legal and administrative framework to facilitate urban commons (Civic eState 2021).

BLENDING THE CONSTITUTIVE AND DECLARATORY APPROACHES

In this chapter we have seen how different cities have adopted different legal approaches. On the basis of the experience supporting urban commons in other Italian and EU contexts and on the basis of the experience of the first three years of implementation, including with the pacts of collaboration funded through the UIA initiative Co-City Turin project, the city Council of Turin repealed the "Regulation on the collaboration between citizens and the city for the care, shared management and regeneration of the urban commons" and passed a new version of the "Regulation on governing the urban commons in the City of Turin" (hereinafter called the New Turin Regulation). The new version was approved on December 2, 2019, and entered into force as Regulation no. 391 on January 16, 2020. The New Turin Regulation emerges out of the recognition by the city of Turin that they needed a broader toolkit to create a different set of governance arrangements to overcome procedural challenges arising from the pacts of collaboration.

The New Turin Regulation importantly introduces two new institutions. One is the Register of the Guarantors, designed to gather experts and city residents with experiences collaborating and trying to construct urban commons. The second is the Permanent Council, a body composed of eleven members from the Register of Guarantors appointed by the city council every three years. The Permanent Council has key consultative and arbitration functions over disputes arising during the implementation phase of the *civic deals* or in the selection of proposals. It serves as a bridge between the relevant community and the city council. The Permanent Council promotes public debates and constitutes a venue for discussion on urban commons co-governance. It also has a role in the evaluation of the activity and results of urban commons co-governance activities. Anyone can address the Permanent Council to protect or safeguard an urban commons. Hence, this body has a general duty to promote urban commons and to mediate between interested parties around those resources.

From a legal standpoint, the first innovation of the New Turin Regulation is that it frames the approach to co-governance of shared urban assets within one general legal category, the *civic deal*, to institutionalize the recognition and grant rights of collective use, management, stewardship, and ownership. (Albanese and Michelazzo 2020, 253). The other innovation of the New Turin Regulation that it embraces two types of urban commons co-governance: (1) shared governance, and (2) self-governance. Each of these two forms of governance anticipates specific legal tools that can be adopted to institutionalize collective action by a variety of participants. Building on the lessons of cities like Naples, and the Bologna experience, the New Turin Regulation avoids limiting the legal tools and gives rise to different options for the governance of urban commons.

For the shared governance element, the legal tool offered by the Turn Regulation is the pact of collaboration. The pact can result from either a city initiative (implying a public consultation process) or direct initiative of city inhabitants (defined by the Regulation as "civic subjects"). The regulation also defines the boundaries and limits of collaboration for shared governance. Whether the initiative comes from the city or civic subject, a responsible manager gathers and assesses the proposals along with the Technical Board. Each project undergoes a co-planning phase that brings together participants, the local government unit, and the Technical Board for collaboration. Another innovative aspect and the heart of the New Turin Regulation is its establishment of three innovative legal tools for self-governance of urban commons: provisions for Civic and Collective Urban Uses, Civic Collective Management, and an Urban Commons Foundation (Albanese and Michelazzo 2020, 257–283). The civic and collective urban use provision is a type of collective right to use civic assets. The urban community of reference uses, manages, and takes care of the urban commons in accordance with the principles of the regulation, such as inclusion and accessibility, guaranteed by a Self-Governance Charter democratically written and approved, similarly to the Naples example previously mentioned. The initiative is overseen by the local government, which ensures consistency with the principles of the regulation and the technical feasibility of the Charter before approving it, therefore closing the civic deal. This form of self-governance ensures that the provision of a municipal good or property is connected to the community of reference. The administration continues to monitor use of shared urban resources or commons to ensure consistency with public use principles. In other words, the community of reference has operational autonomy of the resource, but the property and legal title stays with the local administration. As such, the responsibility for urban commons is shared between the public administration and the civic subjects.

The civic collective management goes a step further in terms of self-governance. First, the contours of the civic deal for the management of urban commons are framed by the community of reference. When the local government approves the civic collective management for a municipal asset, however, the community of reference must manage it in accordance with the regulation principles through the self-governance Charter and the community of reference assumes all relevant liabilities, even though the property remains with the city administration. Thus, the collective civic management is not an expression of the administration's will. It is rather a manifestation of collective autonomy, provided that the principles of openness and democratic accessibility of the space are respected, and that the administration recognizes it, closing a civic deal of which they do not directly determine the content.

We discuss the last self-governance legal tool, the Urban Commons Foundation, in the next chapter for its relevance from a financial standpoint, which is one of the most critical aspects for the long-term sustainability of the urban commons once legally recognized. Through the Urban Commons Foundation, the city can transfer one or more of its assets to a Foundation established for the sole purpose of managing urban commons in the general, public, common interest.

In this chapter, we have reviewed various examples of cities that are experimenting with different legal tools that enable and empower urban communities to utilize city resources and infrastructure to collectively create and sustain different forms of urban commons. The legal pluralism that emerges from the various cases has been categorized into two main approaches: constitutive (adopted, for example, by Bologna) and declaratory (adopted, for example, by Naples). Those approaches are implemented through a variety of legal tools ranging from collaboration pacts or agreements to civic use regulations that allow the private use of a public asset. These legal tools, however, are crafted for the specificity of each case and adapted to the local political and economic context. The next chapter builds on these examples and introduces additional factors that determine how these policies can evolve and be replicated across contexts. We examine in particular how important the creation of appropriate forms of urban partnerships is to facilitate collaboration between the public and community or the public, private, and community actors and to ensure effective implementation of the legal tools discussed in this chapter. The next chapter addresses the question left open by these first case studies, namely, whether legal tools are enough to promote the reduction of power asymmetries in the city and to enable urban communities to leverage legal and financial mechanisms that ensure the sustainability of urban commons and the survival and flourishing of the communities relying on them.

4

URBAN CO-GOVERNANCE

In chapter 3 we explored some of the first public policies and regulatory instruments that have enabled city residents to utilize resources and infrastructure collaboratively or collectively throughout the city. These policies and regulations reflect a commons-based approach to governance at the level of the city that fosters collaboration and cooperation as a methodological tool through which heterogeneous groups of individuals and institutions co-create or co-govern the city or parts of the city as a common resource. At the level of a common pool resource, *engaged citizens* become problem solvers and resource managers, able to cooperate and make strategic decisions about common assets with other urban stakeholders. To successfully realize these policies, local governments embrace a dynamic and adaptive approach to public administration and urban collective welfare.

We saw that these policies can embody two different regulatory approaches, a declaratory and a constitutive approach, depending on whether the local regulator intends to recognize a pre-existing legal entitlement or instead grant the right of city inhabitants to self-organize and to govern critical urban services and infrastructure. The policies described in chapter 3 put in place new public processes and spaces that bring together citizens and other stakeholders in informal settings to change the democratic and economic functioning of the city and to create a city that better functions according to the needs of all its citizens. The creation of shared, common resources in the city is a way to improve the quality of life of residents and to support new local economies, namely, collaborative economies that provide services to satisfy emerging unmet economic and social needs of the neighborhood. Shared urban resources foster social and economic inclusion when they are used as places or means for people to learn skills, obtain access to job opportunities, socialize, and access social services that increase economic empowerment. Yet, we know that it takes more than new policies and laws to change the political economy of cities, including the vast inequality within them. For shared common resources to realize these functions, we must attend to the messy business of how these institutions are created and stewarded, making sure that they serve a distributive function to meet the needs of those with fewer resources.

This chapter builds on the concept of urban co-governance to explain three tenets: (1) communities are a necessary but not a self-sufficient actor; multistakeholder cooperation is essential for the long-term durability and effectiveness of urban commons; (2) as a consequence of tenet 1, legal recognition and legal tools (e.g., pacts of collaboration or civic uses) recognizing or granting governance rights to communities are not enough; what is required are policies and programs that provide a set of enabling conditions that structure complex forms of cooperation in the form of public-community and public-private-community partnerships; and (3) in addition to legal tools, creating a set of enabling conditions requires institutional, learning, capacity-building, digital, and financial tools.

IT IS NOT JUST ABOUT THE COMMUNITY

In this chapter, we explore what we have termed *urban co-governance* reflected in policies and in settings in which communities interact with the state and other actors to co-create and co-govern urban resources such as land, buildings, and even utilities and wireless networks. These are settings that contribute as much as policies and regulations can to enable engaged citizens in the pursuit of their goal to implement arrangements in which citizens are *governing* and not simply being *governed*.

The first aspect to highlight about urban co-governance is that it embraces the role of the enabling state, in which city officials and staff are tasked to assist by providing resources and technical guidance to help create the conditions for co-governance, sometimes in the form of public-public and public-community partnerships. To imagine the local government as a facilitator, it is necessary to move away from the Leviathan or the Gargantua state, which both represent a centralized agency that manages all citywide goods and services such as transit and utilities, to design an institutional system without a dominant center. This kind of system involves other actors in decision making and administrative implementation processes, considering such actors to be peer co-workers or co-designers. This system is characterized by a move away from a vertically (top-down) oriented world to a horizontally organized one in which the state, citizens, and a variety of other actors collaborate and take responsibility for common resources.

Second, urban co-governance creates a system that at its core redistributes decision-making power and influence away from the center and toward a network of engaged urban actors. The co-governance model that we embrace takes as a starting point the active involvement and participation of citizens in the management and governance of urban resources to support the livelihood and well-being of their communities. However, co-governance implies the involvement of other actors, including public authorities, private enterprises, civil society organizations (NGOs), and knowledge institutions. The only question is how to think about and conceptualize their involvement. Our urban co-governance framework is a multistakeholder scheme whereby local communities emerge as key actors and partner with at least one of four other actors to co-produce and co-govern urban, environmental, cultural, knowledge, and digital commons. We have in previous work referred to this co-governance model as a *quintuple helix* system involving five key actors: (1) city inhabitants taking action for the greater good, including *commoners* and activists on the urban commons, social and urban innovators, city makers, and organized or informal local communities; (2) public authorities; (3) private economic actors, including national and local businesses or small and medium enterprises interested in sustainable development and deeply rooted in their regional specialization, social businesses, and neighborhood or district-level businesses; (4) civil society organizations and NGOs; and (5) knowledge institutions, including schools, universities, research centers, cultural centers, and public, private, and civic libraries.

The triple helix concept has been utilized in innovation research and policy studies to mark a shift from an industry-government dyad characterizing the *industrial society* to a triadic relationship between university, industry, and government in the *knowledge society* (Leydesdorff and Etzkowitz 1998; Barca 2009). The basic idea is that the potential for innovation and economic development in a knowledge society lies in the hybridization of elements from university, industry, and government to generate new institutional and social formats for the production, transfer, and application of knowledge. "The interactions between the three strands of the 'helix' creates the unique and distinctive characteristics of an innovation system . . . at either a national or regional level" (Harding 2009, 142). This *triple helix* model of innovation is characterized by a university-industry-government relationship that promotes the role of university in technological and economic development processes (Etzkowitz 1993; Etzkowitz and Leydesdorff 1995).

This model foresees a particularly active role for universities and other scientific institutions, as *entrepreneurial* and *engaged* institutions (Etzkowitz 2003; Levine 2007). The role of university and other members of the scientific community has changed over the decades, thanks to the increasing intertwining of science and technology that is pushing scientists, including social scientists, to be more aware of the implications and impacts of their work and therefore their usefulness as actors of societal change. Knowledge production is increasingly socially distributed. Increased access to higher education has produced the largest number of people in history equipped with research methods that they carry into their own lives and therefore expand and distribute knowledge production (Gibbons 1994).

The educational approach adopted today by many universities is much more engaged with society and communities in which those universities are situated. An engaged educational approach underscores the importance of connecting research, learning, engagement, and leadership to provide students a spectrum of practical applications within and across disciplines (Fung 2016, 2017). This approach might be viewed as the so-called third mission of knowledge institutions—production of an impact on society—beyond simply providing educational services or carrying out research activities that observe and explain social, economic, and natural phenomena. It implies action, and it requires actionable scientists. This has led to the possible emergence of a *quadruple helix* model of innovation, in which the fourth helix is represented by civil society organizations (Etzkowitz and Leydesdorff 2000; Carayannis and Campbell 2009; Carayannis, Barth and Campbell 2012; EC 2016).

Similarly, some are arguing that universities should be engaging in local policy and local economic development, not only for their own interest in becoming key territorial players but also to expand their role as actors of economic growth and social progress beyond patent creation and product commercialization (Breznitz 2012). Thus, a quintuple helix model adds to societal engagement the crucial issues of social and economic inclusion of low-income individuals or communities and the daunting challenge of environmental and economic self-sustainability of local ecosystems (Carayannis et al. 2012; Iaione 2016; Foster and Iaione 2019; Enas 2019).

However, as showcased by the Horizon 2020 research and innovation EUARENAS.eu project, which applies the co-city approach in four pilot cities Reggio Emilia, Budapest, Voru, Gdansk, the quintuple helix model needs to be adapted to specific cities' contexts and leave the door open for the emergence of new actors or the reinterpretation of existing actors. The lesson learnt so far through the project is the need to adopt an "n-tuple approach" (Leydesdorff 2012) and start including the role of media and future generations as well as other normally voiceless actors such as kids, endangered species, or ecosystems (Nagy 2009). This is to say that the structure of helices should be interpreted as more open, evolutionary, and flexible in terms of the number and type of actors to be considered or involved.

On the basis of our empirical and experiential observation, we embrace—and more thoroughly develop this argument in the next chapter—this quintuple helix model to envision, on one side, a bridging role of universities between public and private authorities, and on the other side, local anchor social organizations (and informally organized city inhabitants and communities). This model requires significant engagement with each of the actors involved. It is in line with economist Mariana Mazzucato's view that cities can be important drivers of innovation and a key lever for the implementation of missions-driven research and innovation policies. The model and her approach promote the notion that public authorities encourage bottom-up solutions and experimentation, participation across different actors, stronger civic engagement, and new forms of multistakeholder partnerships for the co-design and co-creation of technological innovations oriented toward solving societal challenges that technological innovations alone cannot solve (Mazzucato 2018b). Within this framework, universities may become critical actors involved with the effort to serve their communities and generate local coalitions for sustainable development, applying social scientific and monitoring/impact evaluation methodologies (Iaione and Cannavò 2015; Iaione 2016).

A final contribution to our model of co-governance to build a bridge between science and society is citizen science theory. The literature on citizen science urges researchers to provide greater access to and the open sharing of research results and findings. Some have specifically argued for *city science*—a greater involvement of scientific research to address governance challenges related to urban assets, services, or infrastructure and to help shape relevant policies (Acuto 2018; JRC Science Hub Communities 2020a; Nevejan 2020). City science advocates for cities to rely on scientific methods and empirical evidence to inform policies and to develop solutions to urban challenges requiring high levels of technology and the use of big data. Citizen science initiatives can complement a city science approach, bringing about the "coproduction of scientific knowledge" (Berti Suman and Pierce 2018). By linking city science and citizen science, local governments can more deeply engage city residents and envision a more robust role for them as not only participants in but also contributors to the process of urban innovation, alongside the private and civic sectors.

Citizen science in the urban context is often directed toward the development of solutions to cope with climate change and pandemics that require cities to mitigate and adapt to these kinds of existential and daunting challenges. Citizen production of data through both analog and digital tools can inform and shape better functioning and more equitable cities through collaboration between scientists/universities and citizens as well as through technologies such as geographic information systems (GIS) (Berti Suman and van Geenhuizen 2020). Citizen *sensing*, for example, utilizes analog and digital mapping and data collection tools through which city inhabitants or communities, supported by researchers and community organizers, produce and analyze data on environmental issues such as air quality, climate change, or natural resources monitoring (Woods et al. 2018). Other applications of citizen science include data collection to improve earthquake or fire preparedness (Verrucci et al. 2016).

An important example of city science in action is the City Science Initiative of the Joint Research Center of the European Commission (JRC Science Hub Communities 2020a, 2020b), which has built a network of City Science Offices (CSOs) in cities such as Amsterdam and Reggio Emilia. These CSOs are internal administrative structures that take a scientific approach to urban policy making and that apply that approach to the carrying out of pilot experimentations through collaborations between city governments, researchers, and various aggregations of urban actors, formal and informal (Nevejan 2020).

Embedding city science and citizen science in our urban co-governance approach is a key part of the basis for urban experimentalism that is required, we believe, to generate collective benefits for city residents and to truly democratize the local economy. Citizens cannot act alone. Universities, research centers, and other members of the scientific community are powerful allies and can, in cooperation with public authorities and other social and economic actors, act as innovation brokers orchestrating processes of social and economic development at the neighborhood or citywide level by providing human and technical resources, transferring knowledge, convening actors, and stimulating them to work together in order to define common solutions to common challenges. Building on the idea of the helix, our co-governance model reflects the evolution of the practice and integrates the literature on, on one side, innovation ecosystems, engaged universities, and city science, and on the other side, participatory or deliberative democracy, governance of common pool resources, and citizen science.

This co-governance model is currently being experimented with in many cities with which we work, as well as in many of the projects that we have identified and mapped as part of our co-cities project mentioned in the introduction. All the actors in an urban co-governance regime can contribute to the pool of ideas, resources, and activities to support constructing and governing urban commons. The role of the state in this helix is to provide the necessary tools (including appropriate public policies packaged as collaborative devices) and sometimes to connect several networks of actors around the city. The institutional settings in which urban co-governance can take place are places of networking, that is, connecting and coordinating different and autonomous actions for shared goals. The actions, reactions, collaboration, and networking of others in the ecosystem are independent and free but nested within a local governance ecosystem that facilitates the creation and sustainability of urban commons, which are collectively governed shared urban resources.

It bears mention that fostering collaboration between local authorities, communities, and key stakeholders is becoming enshrined in policies at the local, national, regional, and global levels. The New Urban Agenda adopted by the United Nations and the Urban Agenda for the EU, for instance, both recognize the importance for cities to foster multistakeholder collaboration as a means of creating sustainable development and of ensuring a just economic and ecological transition. The framework of urban co-governance that we embrace, including multistakeholder partnerships that center the most vulnerable communities within them, resonates with these international policies at the local and sublocal levels. For instance, Goal 17 of the 2030 Agenda for Sustainable Development of the United Nation, a subset of Sustainable Development Goals, highlights the need to "enhance the global partnership for sustainable development, complemented by multistakeholder partnerships that mobilize and share knowledge, expertise, technology and financial resources, to support the achievement of the sustainable development goals in all countries, in particular developing countries" (17.6) and to "encourage and promote effective public, public-private and civil society partnerships, building on the experience and resourcing strategies of partnerships" (17.7). Similarly, the New Urban Agenda (NUA) of Habitat III, which contributes to the implementation and localization of the 2030 Agenda for sustainable development, promotes "the systematic use of multi-stakeholder partnerships in urban development processes, as appropriate, establishing clear and transparent policies, financial and administrative frameworks and procedures, as well as planning guidelines for multi-stakeholder partnerships." The Urban Agenda for the EU similarly calls for a recognition of "the potential of civil society to co-create innovative solutions to urban challenges, which can contribute to public policy making at all levels of government and strengthen democracy in the EU" (Pact of Amsterdam 2016).

FROM GOVERNMENT TO GOVERNANCE: PUBLIC-PRIVATE-SCIENCE-SOCIAL-COMMUNITY PARTNERSHIPS

Taken from a longer view, one way to understand our urban co-governance approach as well as the legal reforms examined in the chapter 3 is that they represent a move from "government to governance" and from "managerialism to entrepreneurialism" (Harvey 1989). In other words, a move away from a hierarchical and top-down management of public goods and services (government) toward a decentralization of decision-making, recognizing that various actors, decision makers, and institutions can shape policies and deliver goods and services (governance). The rise of this conception of governance ushered in during the 1980s the predominance of public-private partnerships (PPPs) and more generally the extensive use of forms of negotiated decision making between public and private actors (Freeman 2000). This form of governance, at the national and local level, altered the state's role in and relationship to markets. It is marked by the retreat of the state from the direct provision of public goods and critical services. In the urban context, these partnerships were employed to revitalize neighborhoods through programs like urban enterprise zones (now opportunity zones) and the provision of services such as waste management, energy and water distribution, and transportation. Instead of being directly delivered by the public authority, these services were outsourced with an often-limited monitoring role left to public authorities.

As we noted in chapter 2, business improvement districts (BIDs) and park conservancies are often structured as public-private partnerships enabling the private sector to provide for and manage urban commons—that is, streets, sidewalks, parks, and playgrounds—while maintaining them as public goods. Proponents of public-private partnerships often note their advantages in avoiding the costs and inefficiencies associated with centralized public authorities (the Gargantua). But PPPs have also shown several limits including the exclusion of local community participation, and they run the risk of paying less attention to local needs and forgoing community support that could be critical for successful implementation (Harman et al. 2015).

Some refer to the early phase of the shift from government to governance and its embrace of PPPs as "urban neoliberalism," which is characterized by the privatization of public services and the commodification of urban space (Rossi and Vanolo 2015). The emergence of efforts to create urban commons, which are rooted in the kind of urban co-governance approach that we have outlined, is one response or pushback to urban neoliberalism. The policies described in the chapter 3 as well as this chapter might also be viewed as part of the effort to disrupt the continuous slide of urban governance into neoliberalism. These policies are notable for *decentering* the private sector in urban governance and *centering* residents and communities in the delivery and management of public goods and shared urban resources.

One attempt at this centering of residents and communities is to integrate a fourth P, for *people*, into the traditional PPP, to form *public-private-people partnerships* (4Ps). 4Ps are arrangements, formal or informal, developed between three categories of actors: public entities, private companies, and city residents. They have been theorized in relation to a variety of projects from real estate service delivery to city resilience in crisis management or smart cities projects (Marana, Labaka, and Sarriegi 2018; Irazábal and Jirón 2021). The distinction that some scholars make between 3Ps and 4Ps is the focus on city residents, noting the difference between public sector–people relations and private sector–people relations (Majamaa 2008, 60). Successful 4Ps can be challenging to put together but they improve the legitimacy of processes because they increase access to information and create opportunities for inclusive participation of all actors affected, ultimately ensuring a successful partnership (Marana, Labaka, and Sarriegi 2018, 46).

A slightly different and perhaps more layered approach to 4Ps is the community-centered partnership that elevates community members and leaders to play critical roles. Such *community-based public-private partnerships* (CBP3s) have been cited as potentially successful in the delivery of infrastructure goods and services, such as water, and more accountable to diverse urban communities. One powerful example of a public-private-community partnership comes out of a case study from Manila,

Philippines, involving the management and delivery of water to the urban poor, including access and availability to safer water and a reduction in costs to households (David and Inocencio 2001). The principal partners were the public water utility and local government, two private concessionaires of the water utility, and local associations and nongovernmental organizations. Participation of the different parties ranged from the labor required to mobilize the community and to build capacity among its members, to negotiating agreements between the water utility and the private concessionaires, to the management of a mini–water distribution system. The result was that poor communities including informal settlements and slums were granted improved access to water through different modalities, including community-managed water connection, privately managed water distribution, and individual household connection.

Some scholars studying these kinds of 4Ps have identified examples in which private operators specifically and intentionally set up these partnerships to serve low-income and poor communities (Franceys and Weitz 2003). These scholars have noted that the role of communities is that of partners on equal footing. They note with caution, however, that the Manila case just discussed limits the role of the community to the provision of some outsourced services and labor. Other scholars have spotted similar problems in 4Ps in Latin American cities (Irazábal 2016). Similarly, in other contexts multistakeholder partnerships might be problematic if they simply replicate asymmetries in the concentration of civic power and influence that already exist in urban environments. The risk is that high-capacity, high-capital private actors approach local governments and local communities to partner and share the benefits and responsibilities of a partnership, but once the partnership is set in motion the interests of the community tend to be overwhelmed by those of the most powerful member(s) of the partnership, often the private sector. These experiences suggest that crafting partnerships, either bilateral or multilateral, for urban co-governance or sharing of shared assets and service delivery requires acknowledging the complexity of actors' interaction (Le Feuvre 2016).

Another attempt to intentionally involve or center residents and communities as co-producers of vital public services (e.g., childcare, elderly care, education, and food supply) is through the creation of bilateral partnerships without the private sector. So called *public-public, public-citizen,* or *public-community* partnerships (PCPs) can involve residents, neighborhood committees, NGOs, and others in the civic sector, together with the local government, through a variety of organizational structures and arrangements. In one definition of a PCP, it consists of "an organizational form of the conjoint production of public services by municipalities and their citizens based on co-operative principles" (Lang et al. 2013). The researchers who offered this definition undertook a study that compared five different examples of cooperative public-citizen partnerships in Austria and Germany and discovered two different "diametrical starting" points for PCPs, depending on who initiated the project. Those initialized by the local government are labeled *Top-Down-PCPs*, and those that arose from community-based groups are labeled *Bottom-Up-PCPs*. In a top-down approach, the government sets up a cooperative PCP and then invites citizens to participate. In a bottom-up approach, the municipality follows the lead of the community group and provides support without taking over the project or initiative (Lang et al. 2013). Cooperative PCP arrangements, in both the bottom-up and top-down approaches, represent a balancing act between dependency through public funding and organizational autonomy, and their long-term sustainability may depend on their ability to mobilize and sustain reciprocal resources (volunteering and donations) to support public service delivery. Another risk identified in the referenced study is that these PCPs can, over time, turn into bureaucratic organizations or corporate businesses.

We embrace a different kind of public-community and public-private-community partnerships—*public-private-science-social-community partnerships* or simply 5P. Expanding on the notion of pro-poor public-private partnerships (Sovacool 2013), 5P partnerships refer to legal, and even more importantly, economic arrangements in which (1) communities are the main partners as the only true holders of stewardship of local or urban ecosystems; (2) civil society organizations and science or knowledge institutions support and coalesce with local communities to negotiate on an equal footing with public and private actors; and (3) the social, science, and community actors are *shareholders* and not just *stakeholders*, which implies that public and private actors share the value, resources, and wealth produced by these partnerships with them. These partnerships are therefore designed to enable resource pooling and cooperation between at least five possible categories of actors: the key player—communities, commoners, innovators, future generations or more generally the unorganized public—and four other main actors—public authorities, businesses, civil society organizations, and science or knowledge institutions—of the urban commons "quintuple helix governance" (Iaione and Cannavò 2015).

As we have mentioned, a co-city is based on polycentric governance of a variety of urban resources such as environmental, cultural, knowledge, and digital assets that are co-managed through contractual or institutionalized partnerships in which communities and their collective organizations become a central partner not only by sharing political or governance powers but also by capturing the economic value produced by these partnerships (Foster and Iaione 2019). These partnerships are normally established with three main aims: fostering the community role in urban welfare provision, spurring collaborative economies as a driver of local economic self-development, and promoting inclusive urban regeneration of blighted areas of abandoned urban real estate (Patti and Polyák 2017).

Not every partnership foresees an equal role of all the partners. There are loosely coupled legal, economic, and institutional arrangements. The most important feature they need to carry is to be structured in a way that communities drive the process and perceive directly the economic gains of the partnerships. First, members of these communities contribute with labor and receive in exchange due and proportionate outputs, however those are measured. Second, the community captures part of the value produced (Lazonick and Mazzucato 2013; Mazzucato et al. 2019) by specific contractual and business arrangements that make sure that all or a large part of the revenues produced are redirected toward the local community, normally through some special-purpose vehicle (e.g., community co-ops, community land trusts, or participatory foundations) that is designed and set up to reinvest any economic surplus into community welfare initiatives, neighborhood rejuvenation activities, and other local community social and economic development activities.

Public authorities act as institutional platforms and play a quintessential enabling role in creating and sustaining the 5Ps implied by the co-city. They can create the space to convene and connect actors, they hold information and data that is crucial for the elaboration and rollout of co-governance projects, and they can provide seed money as well as other tangible and intangible resources to pave the way for collectivities to engage and self-organize. Against the vision that has dominated since the 1980s, advocating for a massive withdrawal of the state, public authorities are today asked to be more entrepreneurial by nurturing and steering the knowledge economy through missions-oriented spending (Lazonick and Mazzucato 2013) or, better, creating more experimentalist incentives to test innovative solutions through forms of multistakeholder cooperation designed to enable local economies and ecosystems to test new technological solutions or create new economic activities by adopting regulatory measures such as regulatory sandboxes to make sure innovation is not hampered by the existing regulatory framework or sunset clauses to prevent resources from being blocked in unproductive activities. This regulatory approach implies a culture of collaborative decision making and constant impact monitoring and evaluation (Rodrik 2004; Sabel 2004; De Búrca, Keohane, and Sabel 2014; Ranchordás 2015; Iaione 2016; Finck, Ranchordás 2016). Institutional capacity or political leadership may be different from one city to another as well as from one neighborhood to another. That is why it is important to have the third sector and science or knowledge institutions also involved in these projects. They can compensate for possible weaknesses of the public sector and still be actors animated by general-interest motivations.

Local anchor institutions such as schools, universities, research centers, libraries, local branches of regional and national NGOs, long-standing community and faith-based organizations, and more generally third sector organizations operating in the neighborhood, are typically involved in information, education, and communication campaigns as well as in community mobilization, acting as social intermediaries between the private operator and the community. Science or knowledge institutions, as much as NGOs can also assist with the provision of services and materials necessary to the rollout of the project.

What is truly unique about our urban co-governance approach is the critical role that knowledge institutions can play in this 5P model that is overlooked by the existing literature on more inclusive forms of PPPs. For instance, knowledge institutions play a significant role in the capacity-building process, providing advisory and skill development services to local communities (Bina et al. 2015). One example is the Public Collaboration

Lab established in London, a social innovation lab that was developed and tested through an action research partnership between the London borough council and the Design for Social Innovation towards Sustainability Lab of the University of the Arts London (UAL DESIS Lab) (Thorpe and Rhodes 2018). Other examples entail more direct engagement by knowledge institutions—they might decide to be one of the main partners of an urban co-governance arrangement, using their most advanced research units to enhance the educational and advisory functions through involving students and deploying their researchers to augment the innovation brought to the partnership (Iaione and Cannavò 2015). This direct engagement normally entails different technical-capacity-building skills. First, their role in these partnerships could include applied research projects based on their theoretical research that serve as prefeasibility studies for these complex operations. Second, these institutions can operate as innovation brokers and create bridges between other actors to possibly compensate for the lack of local institutional capacity or leadership. Finally, they can pitch these operations to long-term financial investors and private operators willing to invest in or experiment with new processes, products, and solutions (Iaione 2019a).

We are aware that the role of the private sector might be controversial. Its role within a PPP scheme has not always been respectful of the public and the social purposes of projects involving delivery of a public service or providing public infrastructure. We refer to the private sector, in the main, as the local commercial or retail sector (e.g., shops, restaurants, and other commercial and industrial activities located where the urban co-governance operation is to be implemented). This sector includes private operators active in the technological innovation and green economy sectors who want to co-create and co-produce new solutions or new products with local communities. They also include neighborhood-based, highly local businesses that are rooted in their communities and are more committed to leveraging their economic power toward community and public ends.

On the basis of our own observations and field experience, we appreciate that the private sector is often an essential actor if urban co-governance is to scale and allow the co-governance of large, complex resources and infrastructure. Communities' interest can be safeguarded by carefully structuring the governance of these partnerships, in our view, if public, social, and knowledge actors are able to assist communities in designing and negotiating the terms of these economic and legal arrangements. There is indeed a growing attention to UN Sustainable Development Goals by companies that declare themselves to be purpose-driven or impact-oriented, embedding social and environmental objectives into their organizational mission (Harrison et al. 2020). This of course raises concerns of green washing, which in turn makes room for the role of non-state actors (Dahlmann et al. 2020).

As we discuss in the next section, there is no single mode of co-governance that will work in every setting. However, what we observe in many cities is an attempt to create spaces for experimenting within a particular local context with how to bring together different actors in settings that are open and transparent and can be designed to center the most vulnerable and marginal populations in the co-governance project. These co-governing processes and institutional architectures are new and could be complex to design. They certainly contain their own dangers. For instance, they can break down when local (or sublocal) factions no longer agree with the governance process in which they are involved or no longer agree with the goals or plans designed for a neighborhood or for a particular local good. Nevertheless, these informal spaces are critical sites for developing a process for the self-organization or the co-design of shared spaces and the co-production of shared urban resources and community services.

SCALING THE LADDER OF CITIZEN PARTICIPATION

A variety of sublocal and participatory institutions and practices have already reconfigured the relationship between central governing authorities and urban residents. Many cities, for example, enable a variety of forms of participatory or collaborative governance through the creation of distributed sublocal bodies, largely advisory, that provide opportunities for direct community input into local government decision making about goods and services in their communities and facilitate interaction with various stakeholders. Neighborhood or borough councils and community boards, for example, advise local governments on land-use planning and neighborhood development as well as the city budget process and local service delivery. New York City, for example, has fifty-nine *community boards*—at least a dozen in each borough (except Staten Island, which has only a handful)—consisting of residents with significant interest in their communities appointed by their borough presidents.

Community boards were created in the 1960s on the heels of the *urban renewal* era as part of a larger effort to decentralize city planning and land-use decision making in response to critics like Jane Jacobs, who argued that top-down, uniform planning in cities destroyed the social fabric of communities (Jacobs 1961). Developers and planning czars like the infamous Robert Moses were given excessive powers to raze working class and ethnic minority communities to build new infrastructure and housing, without any real accountability (Rae 2003). Community boards like New York City's give neighborhoods and community members a strong voice in land-use, planning, and zoning decisions that shape their communities, and they create at least the appearance of accountability for development decisions and in empowering neighborhoods and communities. Similar entities exist in other American cities as diverse as Raleigh, North Carolina, Seattle, Washington, and Washington, DC, which has forty Advisory Neighborhood Commissions.

Advisory boards or commissions, as their name suggests, have no binding authority in the city's centralized decision-making process. Their input and recommendations are not binding on public officials, although some cities like Washington, DC, require their input to be given "great weight." This is not to say that these institutions are without influence on city decision making. There are plenty of examples from cities across the US of individual community boards and commissions providing input that has improved city planning and altered the details of specific infrastructure projects to be more responsive to a community's needs. Nevertheless, in most instances these bodies do not decentralize formal influence or power. Decision-making authority remains quite centralized within city council and city hall, albeit with improved input from communities. Nor are these bodies designed or set up to require co-governance in any meaningful sense with local authorities.

The reality of many (if not most) community boards and commissions is that they are often the site of a power struggle between elected officials, developers, and communities. The strength of a specific advisory board or commission depends on the strength of resources within the community it represents and its ability to influence elected officials. Studies show that the communities that are often on the losing side of the struggle with developers and local officials are those with a lower level of income and wealth and that have comparatively fewer resources to compete in the pluralist marketplace for a voice in city politics (Rogers 1990). Increasingly, scholars of local politics also believe that these sublocal units tend to bias decisions and policy discussions in favor of an unrepresentative group of individuals in communities, such as homeowners and longtime residents, who tend to oppose new construction—particularly new housing—that adds to neighborhood density and aids the development of more affordable housing (Einstein et al. 2018).

The emergence of community benefit agreements (CBAs) in large-scale urban development projects is a potent reflection of the changing relationships between local government, private developers, and urban communities. CBAs are private agreements between developers and communities most impacted by a development project. Pioneered in Los Angeles at the start of this century, CBAs stand outside the normal government process as private, legally binding contracts between community groups and developers (Gross et al. 2005). In these enforceable agreements between community groups and developers, both developers and communities face incentives to participate and negotiate with one another: developers bargain directly with the community to win its backing for the project or, at least, neutralize its opposition, and communities participate out of a desire to mitigate negative development impacts and maximize development benefits (Salkin and Lavine 2008). Projects that are in significant part subsidized by taxpayer funds are now going to finance affordable housing, jobs, environmental, and infrastructure amenities and other "benefits," thus returning some of that money to the public/community.

The limited scale of CBAs, usually growing out of resistance to a large development project, does exclude a high percentage of urban residents who would profit from having a central role in negotiations with developers in many other city projects (Raffol 2012). CBAs also are limited in their ability to stimulate economic growth in low-income communities, given that they exist to extract benefits from a particular development project that otherwise might have resulted in harm to the most vulnerable members of the community. Nevertheless, although some question whether CBAs deliver long-term economic benefits or housing affordability to communities, these are deliverables that in the absence of a CBA, local communities do not believe will be negotiated nor provided for on their behalf in any deals struck by local governments with private developers (De Barbieri 2016).

The *community* in the CBA consists typically of some coalition of place-based NGOs including environmental organizations, labor unions, and neighborhood groups (Marantz 2015). The power of the landmark Los Angeles CBAs, for example, stemmed primarily from the strength of the forces that came together to bargain with developers. Those coalitions united community groups with powerful, well-staffed, citywide progressive organizations, major unions, and strong national environmental organizations. Staff and leaders of these high-capacity organizations led the negotiations, in consultation with community activists (Cummings 2006). Whereas city officials play important background roles in the CBA bargaining process and are sometimes made an additional party to strengthen capacity to monitor and enforce compliance, the CBA process empowers impacted communities to develop their own vision for development within those communities. CBAs represent a step away from the mainstream community economic-development model of heavy dependence on outside capital, which determines what types of facilities can be built and what services provided in low-income communities (Cummings 2001). However, each CBA is differently structured, and properly understanding the balance of power between the different parties is essential in understanding whether the CBA will lead to true benefits to the concerned parties, how much it will improve communities, and how to minimize its dangers and disadvantages (Wolf-Powers 2010).

There is, however, a "new public participation" emerging in cities around the world that places citizens and communities at the center of decisions about city resources, including arguably the most common resource of all in cities—the city budget (Lee et al. 2015). Participatory budgeting practices are much higher up on Sherry Arnstein's iconic "ladder of citizen participation" than neighborhood advisory councils or even CBAs. Arnstein's ladder of citizen participation is a framework for understanding how much power citizens have in any given institutional form (Arnstein 1969). These levels range from very low to maximum delegated power: manipulation, therapy, informing, consultation, placation, partnership, delegated power, and citizen control. Participatory budgeting is emblematic of one of the higher steps on the ladder of citizen participation, beyond consultation and even partnerships. Depending on how it is carried out, participatory budgeting practices arguably represent delegated power over decisions about the allocation of public funds for their communities.

Participatory budgeting (PB) began in Porto Alegre, Brazil, in 1989, adopted by the progressive Workers' Party to help poor citizens and neighborhoods obtain higher levels of budget resources (Wampler 2000). It has since spread like wildfire across cities around the world, including large cities in the US, Europe, and other advanced economies. Participatory budgeting has also spread to places like China, giving public legitimacy and increased citizen input into the budgeting process despite the constraints imposed by China's contemporary party-state system and its monopoly on political power even at the local level (Wu and Wang 2012). In western democracies, it is safe to say the process has evolved a lot more quickly. Most participatory budgeting processes in the US and Europe have focused on giving residents more power in the allocation of capital works or infrastructure dollars (Baiocchi and Ganuza 2014; Lerner and Secondo 2012). In most places, the process takes a recognizable form: a representative resident committee creates a plan and rules for the process in partnership with government officials; residents share and discuss ideas for projects in their communities; members of the community become "budget delegates" and turn the community's ideas into actual proposals; community residents vote on the proposals that the delegates have created and finally the government funds and helps to implement the winning proposals and, together with residents, tracks and monitors their implementation (Participatory Budgeting Project n.d.).

One of the first implementations of participatory budgeting in the United States was in New York City in 2012. In its first year of implementation, eight thousand community members were able to participate in the decisions surrounding the allocation of around $6 million in New York City (Lerner and Secondo 2012). By 2015, community engagement had increased to over fifty-one thousand community members weighing in on how the city should spend $32 million, and the number of districts using PB increased from four to twenty-seven. In 2019, community members in New York City had a chance to vote on how to spend over $35 million on capital projects such as physical infrastructure (Spivack 2019). Community members were able to submit project ideas on an "idea collection map" that was later shared with budget delegates, who turned the feasible ideas into proposals that the members could later vote on. To qualify to vote, the Council's only requirements are that a community member must be at least eleven years old and live in a district that is taking part in participatory budgeting.

In Chicago, the city delegates around $1.32 million to each council member, leaving it to them to choose whether to delegate the decisions on how that money will be spent to their constituents. Over twenty-six thousand residents have assisted in the decision-making process surrounding the allocation of $31 million in public funds through over 160 public projects (Great Cities Institute 2020). In those wards that do choose to use PB, community members brainstorm ideas, proposals are developed using their ideas, project expositions are held in which the proposals are explained, and then residents vote on the proposals (PB Chicago 2020). The most common project type that community members have chosen to implement involves streets and sidewalks, with a total of 109 projects in many different neighborhoods. There have also been forty parks and environment projects, twenty-five arts and culture projects, eleven bike and transit projects, and ten libraries and school projects (PB Chicago 2020).

Participatory budgeting is a tool for treating the public budget as a common resource and giving communities power and maybe even control over determining the best ways to utilize that resource to meet their needs. In this sense, it resonates with the kind of decentralized, enabling, and polycentric governance system that we detect in some of the policies that we describe in this book and that we embrace as part of our idea of urban co-governance. Designing co-governance processes or institutions to include a wide range of citizens, particularly those most vulnerable to being excluded from decision-making processes because of their social or economic status, is important for managing any common pool resource.

Yet, as the literature on urban planning suggests, even the best participatory or collaborative practices, and especially those that simply devolve planning processes to the sublocal level without offering new tools and resources to enable meaningful involvement by the most vulnerable populations, are prone to domination by economic elites and/or strong or corrupt sublocal leadership (Elwood 2004). Not surprisingly, we see that the promise of participatory budgeting has predictably run up against the peril of entrenched inequalities in participation and influence that beset many democratic processes (Young 2000). Experience with PB in the US, for example, is raising concerns about the ability of the process to reach individuals and communities that are normally underrepresented or marginalized in democratic processes at the city level.

A common critique of participatory budgeting practice is that the "usual suspects," typically white middle- to upper-middle-class homeowners, are the ones to turn up at the meetings and participate in the vote and therefore to control how the money is allocated (Pape and Lim 2019). Chicago's participatory budget process suffered from this fatal flaw, according to a recent study, through its design, which made it "not only inaccessible but also less attractive to the residents whom organizers of the process claimed to be serving" (Pape and Lim 2019, 878). The study found that residents who voted in PB Chicago were more often white, college educated, and from higher-income households relative to both the local population and the politically active residents in Chicago. One of the reasons for these patterns of participation, the authors concluded, was that the "menu budget" for the PB Chicago project over which aldermen have autonomy "is aligned with the interests and agendas of already-privileged residents, and which can therefore be opened up to public decision making without fundamentally challenging the status quo" (Pape and Lim 2019, 878).

The marginalization of the poor and vulnerable from claiming and sharing in the benefits of decentralized or devolved decision-making power need not be the inevitable result of these processes, particularly if the process is intentionally designed toward strong inclusion of those populations and communities. In Porto Alegre, Brazil, for example, the focus on capital works projects meant that participatory budgeting was geared toward the priorities of less privileged residents, particularly those living in irregular housing settlements without basic infrastructure (Pape and Lim 2019). Likewise, in New York City, the "usual suspects" did not dominate the process because of the "larger and more flexible pot of money" that communities had to work with, which allured many different community groups, and because of the city's partnership with a nonprofit organization that focused on organizing low-income women of color (Lerner and Secondo 2012, 6).

One of the lessons from experiments with participatory budgeting, and other forms of decentralized local decision making is that even the highest form of participation and citizen power can fall short of altering the unequal power dynamics, privileges and advantages that often characterize urban geographies that are stratified by class, ethnicity, immigrant status, and race. The challenge for any system of participatory or collaborative true polycentric governance is to avoid replicating the very inequalities and power dynamics they are often set up to address. The best collaborative urban processes, in our view, will intentionally and deeply engage and empower the most vulnerable stakeholders in any co-governance or *partnership* process, arrangement, or agreements.

One notion that maximizes this idea of deep engagement and empowerment is benefit sharing. This notion is rooted in the idea that those who either hold stewardship rights of an essential resource or contribute to the development of any kind of intangible or tangible resource (such as cultural heritage, natural resources, or scientific knowledge) are entitled to benefits connected to its use, reuse, or development (such as for example a share of the intellectual property rights or a portion of the economic profit produced through its use). From a governance perspective this notion can be implemented through benefit-sharing agreements which imply an exchange between local communities or institutions granting access to a particular resource and business operators providing compensation or reward for its use.

In a case described by Wynberg (2004) a benefit-sharing agreement was created in South Africa around the revenues of commercial development of a botanical plant-based product. It involved the patent holder and the indigenous community that was the traditional holder of the knowledge around its use and properties. The parties committed to the preservation of biodiversity in the region. Additionally, the agreement provided that the indigenous community, the San, would receive 6 percent of the royalties and 8 percent of milestone income received by the holder of the patent from the pharmaceutical company commercializing the product.

The money would go into a trust created by the patent holder and the Council of the indigenous community, the South African San Council, destined to be used to improve the quality of life of the people (Wynberg 2004, 863).

Benefit-sharing agreements are highly differentiated across regional areas and types of economies, for example the Arctic or sub-arctic regions and the energy sector (Pierk and Tysiachniouk 2016; Wall and Pelon 2011; Gonzalo and Thimbault 2014). Specific attention is devoted by the scholarship to remote areas in the mining sector. (Tysiachniouk and Petrov 2018). Benefit-sharing agreements recognize that local communities have the right to grant companies a *social license* to operate, SLO (Prno and Slocombe 2012). This social license would guarantee social acceptance of companies' activities provided that binding requirements for resource extraction set forth in the license are respected (Morgera 2016). On the basis of governance and distribution mechanisms, benefit-sharing agreements have been classified in four categories: (1) paternalistic; (2) company-centered social responsibility (CCSR); (3) partnership; and (4) shareholder (Tysiachniouk and Petrov 2018, 30–32). The 5P model falls in principle within the shareholder category, which involves dividends and royalties redistribution to local community members.

Even the most intense form of benefit sharing such as the shareholder mode lends itself to criticisms. Critical aspects include the risk of regulatory capture, pressure on local communities to commodify their own resources, or even worse trigger or exacerbate conflict and individualism in local communities (if benefits target individual beneficiaries instead of the community). In fact, the presence of a possible benefit might nudge individuals to go after the monetary or nonmonetary value embedded in the benefit to be shared and distract them from the need to safeguard the conservation of communities' essential resources and steward their sustainable use. This might ultimately result in disadvantaging minorities, future generations, and endangered species or ecosystems.

Olajide and Lawanson (2021) analyze a Memorandum of Understanding (MoU) signed between the Lagos State Government, Ibeju Lekki Local Government, Lekki Worldwide Investment Limited and the representatives of the communities affected by an urban real estate development project. The MoU provided for resettlement of and compensation for these communities including the recognition of 2.5 percent of capital shares of the investment company and a seat on the board of directors, as well as alternative housing, capacity-building, and social development projects.

According to Olajide and Lawanson, the MoU was not a good deal in safeguarding their economic rights. Communities were able to preserve their homes but lost their farmlands (and consequently their livelihoods), which were acquired by the government and compensated not at their fair market value. Olajide and Lawanson argue that the government did not negotiate in good faith with the objective of maximizing the interest of their communities, but rather acted as an agent having in mind the maximization of the market-oriented operators' interest.

Thus, benefit sharing arrangements if engineered as 5Ps might prevent communities from entering in such bad deals because they imply the simultaneous involvement of social watchdogs and science or knowledge institutions. We argue that the presence of these two other actors might further contribute to increase accountability, transparency, and above all the ability of local institutions and communities to negotiate more equitable, community-driven, collective benefit-sharing agreements (Morgera 2016).

In any case, even 5Ps benefit sharing agreements can overcome the shortcomings of traditional benefit sharing agreements only if they are carefully designed and negotiated in such a way as to effectively enable collective benefit sharing and foresee investments on human capital through reskilling processes, as well as to include real and equal governance and economic rights for the unrepresented or underrepresented interests in local communities.

In order to avoid races to the bottom, 5Ps should therefore be designed as a tool to enable multi-stakeholder cooperation and strike a fair deal for communities; the presence of social, science, and knowledge institutions might be particularly critical to support communities in negotiating better terms with public and private actors. In chapter 5, we offer some thoughts about and examples of how some cities make conscious and intentional efforts to create spaces for similar co-governance arrangements that center the most vulnerable urban populations and communities and empower them in the process of co-producing and co-owning critical urban resources.

ENABLING THE COMMUNITY IN CO-GOVERNANCE

One of the defining features of many of the policies discussed in chapter 3 and similar policies is the embrace of collaborative hubs through which the relationship between local officials, community residents, and various stakeholders is fostered and mediated. These collaborative hubs emerge from and exist outside the local government infrastructure and administrative apparatus, although they work with public actors. One example of how these collaborative hubs manifest is through urban or living labs that operate in a physical or virtual setting, or both, to pool knowledge and resources and to co-design urban solutions (Cossetta and Palumbo 2014; Baccarne 2014). These labs can be established at the neighborhood or city level. Regardless of the geographic focus, these informal spaces emerge as critical sites for developing a process for the self-organization or the co-design of shared spaces and the co-production of community services. At the heart of these living labs is a methodological protocol known as *collaboratory*, conceived in the late 1980s in the field of computer science and later applied to fields such as environmental or energy research (Grimes 2016). The key idea behind the development of collaboratories in scientific research is that knowledge is an activity that is inherently collaborative (Finholt 2002). They operate as a catalyst to foster mutual learning and co-creation (Ostrom and Hess 2007).

The city of Reggio Emilia in Italy, for example, is experimenting with collaboratories at the neighborhood level as an essential tool in reaching citizenship agreements between local authorities and residents pursuant to a Regulation for Citizenship Agreement adopted in 2015 within the *neighborhood-as-a-commons program* (see https://www.comune.re.it/siamoqua). This program was the first policy tool forged to implement this approach and in 2015 initiated collaboratories as co-design sessions that take place in neighborhood social centers to define urban innovation projects with the actors in the neighborhood.

The aim of the neighborhood-as-a-commons policy and the connected regulation on citizenship agreements is to create institutional spaces where public, private, community, and civic representatives can work together to develop community-based institutions and enterprises that offer neighborhood-level goods and services to support vulnerable populations.

Citizenship agreements can involve the reusing, regeneration, constant care, and maintenance of urban spaces. They can also involve the provision of services to small businesses, courses, and workshops on soft and professional skills. and the creation of intangible digital services such as a wireless network. Like the Bologna pacts of collaboration discussed in chapter 3, the scope of each citizenship agreement is designed to reflect the complexity of the project undertaken, including the type of activities engaged in (e.g., care, management, or regeneration of spaces) and the extent of the collaboration (e.g., the number and type of subjects involved, the duration of the collaboration, the advertising measures, the documentation produced, the supervisory capacity of the municipal staff, and the ability of participants to comply with regulatory requirements). The aims of the policy and the regulation, however, go beyond revitalizing urban infrastructure and instead is geared toward supporting "the creation of cooperatives, social enterprises, and start-ups in the social economy sector and the development of activities of and projects of economic, cultural and social value" (Art. 6, par. 5.2 of the Regulation). Between 2015 and 2022 (as of March 2022) thirty-nine citizenship agreements had been signed in neighborhoods across the city (Levi, 2018; Comune di Reggio Emilia n.d.).

Departing from the Bologna regulatory based on collaboration pacts, Reggio Emilia's approach to reaching neighborhood agreements recognizes the need to implant community spaces rooted in each neighborhood that are aimed at improving the capacity of local communities to develop social economy and responsible innovation ideas. A key player in the collaboratory process enabled by the Reggio Emilia regulation is the *neighborhood architect*, a civil servant who assists urban communities in steering the process leading up to a neighborhood agreement. The neighborhood architect is an innovative administrative figure who acts as a broker between the city, neighborhood inhabitants, and other local stakeholders. The skills required for the neighborhood architect are knowledge of the fundamentals of project management in the field of social innovation, communication, and relational network management skills. The neighborhood architect is active in the different phases of the co-design process in negotiating and closing citizenship agreements that convene all the actors around the delivery of new neighborhood-level services and/or the regeneration of urban spaces or buildings (Antonelli, De

Nictolis, and Iaione 2021). At the beginning of the process, the architect works with communities that joined the citizenship laboratory initiated by the city and helps them define a common vision of what they want to improve in the neighborhood. With the support of the city's social innovation central unit, the architect then makes sure that other actors become partners and commit to designing the solution according to the challenge identified by the neighborhood's inhabitants.

The development of collaboratories as a network of community social centers is the means adopted to reach neighborhood agreements considering the different capacities and needs of diverse neighborhoods and communities. They serve as incubation spaces for developing ideas and measuring and monitoring the impact of the new initiatives. These collaborative hubs act also as co-working spaces. For instance, these kinds of hubs can interact with a city-level policy innovation lab, described in connection with the Bologna experiment, to coordinate and monitor the path of experimental civic collaboration projects around the city. The main difference between policy innovation labs and collaborative hubs is where they originate. Collaborative hubs emerge from and exist outside of the local government structure and administrative apparatus, although they work with public actors. They are often launched in the city's neighborhoods and in this way can be part of an ecosystem of distributed spaces that bring into the center of collaborative processes those that are the most disadvantaged in order to most directly address economic and digital inequalities.

To coordinate the effort of these collaboratories, Reggio Emilia devised the ambitious plan of setting up a citywide evidence-based policymaking lab to be managed as a CSO by a scientific operator such as a research center or a university. Its goal will be to constantly convene and involve quintuple helix actors—who belong to the five different categories (local public authorities and agencies, businesses and local entrepreneurs, NGOs and social actors, city residents and informal groups of innovators, and knowledge actors such as schools and universities) and who pool their resources and cooperate to carry out projects to improve the cities' services and infrastructures—to generate new neighborhood-based digital and social innovation solutions enabling free and fair access as much as co-management and co-ownership of social, green, economic, technological, data, and digital urban infrastructures. The proposed solution is centered on the recognition of local communities as equal partners and therefore active designers and managers of science-based solutions aimed to tackle at the neighborhood level the ongoing ecological and digital transition processes.

We mentioned in chapter 3 the case of the *Laboratorio para la ciudad* (Mexico City) as a successful example of a citywide laboratory to solve urban challenges relying upon residents' engagement. Another example of an urban lab or collaboratory, in a very different city, that centers marginal communities and populations within the context of co-designing and co-creating local tech infrastructures and services that serve the needs of those populations, is New York City's NYCx Co-Labs. Launched in 2014 by the Mayor's Office of the Chief Technology Officer (MOCTO), NYCx Co-Labs are collaboratories focused on the neighborhood level and designed to propose and test new solutions to modernize public infrastructure, support community-driven development, and bridge the digital divide in low-income and ethnic minority areas of the city. The Co-Labs' aim is ensuring that the growing economies of low- and moderate-income communities are front and center in the creation of a smart city. At the time of the creation of NYCx, new technologies were already beginning to transform urban life and city services, and numerous reports were projecting a rapidly growing economic opportunity around solutions for improving urban systems (a 2014 report by Frost and Sullivan projected the market for smart city technologies to be valued at $1.6 trillion by 2020). MOCTO was responsive to the Obama White House Office of Science & Technology Policy's request for cities across the US to join an announcement of smart city initiatives helping communities tackle local challenges and improve city services. With that announcement, MOCTO committed to create a series of "neighborhood innovation labs"—now referred to as the NYCx Co-Labs—in underserved neighborhoods to accelerate the deployment of smart city technologies with systematic collaboration with the New Yorkers whose employment, transportation, health, and environmental circumstances would be most affected by the transformed landscape. By co-developing impactful technologies alongside civic technologists, start-ups, tech industry leaders, and city agencies, residents are expected to play an active role in shaping the future of the communities where they live, work, and play.

The neighborhoods that NYCx Co-Labs target, such as Brownsville in eastern Brooklyn, exhibit some of the city's poorest health, social, and economic outcomes. Life expectancy in Brownsville, for example, is eleven years shorter than in lower Manhattan. A predominantly Black and Latino neighborhood, Brownsville is experiencing an overall 34 percent unemployment rate, and not surprisingly, neighborhood young people are experiencing the highest rate of unemployment in Brooklyn at 44 percent. Median household income in Brownsville is just $25,252—roughly half the citywide average. Simultaneously, crime rates are almost twice the citywide average. Brownsville remains significantly behind in the area of digital adoption, with 45 percent of households lacking an internet connection at home—the lowest rate of any neighborhood in New York and a major impediment to educational and professional advancement. As it has in more affluent areas, participation in civic technology and the innovation economy could catalyze significant gains in economic opportunity and mobility in low-income communities across the city, but to date, it has not. Through structured programming, NYCx Co-Labs offers a unique opportunity for residents to develop civic technology skills, collaborate on the strategic identification of community needs, and apply their newfound tech knowledge to co-creating solutions to local problems.

By design, NYCx is a convener of community stakeholders, policy makers, industry leaders, legislators, external advocacy groups, and public interest technologists. Before formally launching co-labs in a new area, MOCTO's team meticulously scans the existing landscape and identifies an anchoring partner organization to initially help recruit a Co-Labs Community Technology Advisory Board (here called the Tech Board) and participants. The Tech Board is composed of a diverse set of local residents and community-based organizations with a focus on one or more of the following: community organizing, local economic development, technology, technology education, public space stewardship and place-making, and creative and cultural production. For example, the Tech Board in Brownsville is made up of pivotal community-based and government organizations working in the neighborhood: Brownsville Community Justice Center, Brownsville Multi-Service Center, the Brownsville Houses Tenants Association, Friends of Brownsville Parks, Brownsville Partnership, Made in Brownsville, Brooklyn Community Board 16, Pitkin Avenue

BID, the Office of NYS Assembly Member Latrice Walker, What about the Children, Bloc Bully IT Solutions, 3 Black Cats Cafe, the Dream Big Innovation Center, the Knowledge House, and Brooklyn Public Library.

Tech Boards are the engines driving the local co-lab and putting the benefits of civic engagement front and center within the local community. The first role of the local Tech Board was to participate in a series of capacity-building workshops, entitled the Neighborhood Innovation Curriculum, which include sessions on technology prototyping, using data for civic applications, and the ethics and risks of technology in the public realm. This initiative is particularly relevant in a moment when cities are becoming increasingly attractive for tech giants as office headquarters. This growing interest could potentially hamper the affordability of rent and cost of living in the area and displace middle-income residents. This is happening, for instance, in Manhattan, specifically on the West Side, which is becoming a tech corridor for four tech giants (Apple, Facebook, Google, and Amazon) that are building offices that are projected to host twenty thousand middle-wage and high-wage workers by 2022.

After collaboratively evaluating proposals with NYCx and sponsoring agencies, the Tech Board designed an engagement strategy for neighborhood residents to participate in co-designing the solution, identifying success metrics, and structuring the launch of the pilot. In the end, the Tech Board plays a pivotal role in the final analysis of the solution's effectiveness, ultimately helping the city to determine which solutions are worth pursuing at a greater scale. Working alongside the Brownsville Tech Board and city agency partners, NYCx Co-Labs issued the first two Co-Lab Challenges in 2017 which requested proposals from local and global technologists and innovators to co-develop and publicly test new solutions in response to the issues/opportunities defined by the multistakeholder collaboration. The first, the Safe and Thriving Nighttime Corridors, is a partnership with New York City's Department of Transportation and the Mayor's Office of Criminal Justice, aiming to enhance the nighttime experience and use of sidewalks and public spaces in Brownsville. The second, Zero Waste in Shared Space, is a partnership with New York City's Department of Sanitation and the New York City Housing Authority that aims to reduce waste and improve recycling in Brownsville Houses, a public housing development shared by more than 1,300 families.

One of the goals of the New York City Co-Labs is for co-lab communities to co-own the systems that they have helped research and develop. Engaging in co-development partnerships with the technology industry and community stakeholders brings to light questions about how to best manage the intellectual-property and technology-transfer implications in a multistakeholder enterprise. As the co-labs program progresses, the team effort is focused on exploring mechanisms to ensure fairness in how the value that is created collaboratively is shared equitably. They intend to explore mechanisms like joint-venture agreements, accelerator-inspired equity models, distributed ledger systems, and community land trust agreements, and their potential applicability to our programs.

The first co-lab in Brownsville has also created significant demand to scale the pilots to new neighborhoods including Harlem, Queens, Staten Island, and the Bronx. As Jose Serrano-McClain, one of the founding staff of NYCx Co-Labs, told us, the program and its expansion confronts "the reality that our city's racially and economically marginalized communities are at the edge of a precipice that could represent unprecedented disenfranchisement in the age of big data and Smart Cities. As a matter of economic and social justice, we must create opportunities for a new class of civic leaders and solutions to emerge. It is imperative that we aggressively advance tech equity, participation, and ownership in this age of data-driven urbanism. Our work attempts to create pathways for individuals from low-income neighborhoods to become civic leaders through engagement with the technology and the innovation economy that is rapidly changing their communities."

Another important community enabling factor is the use of technology as a tool to support the public-community collaboration. The co-city experiment required the city of Turin to implement a platform based on blockchain technology to function as a civic social network, the so-called First Life platform. On First Life, city inhabitants and other actors can map urban commons, access information about them, reach out to other city inhabitants and intracity networks that might be interested in working on a project for a pact of collaboration, to create groups and collaborate. More sophisticated digital tools that could further enable collective action could entail the use of blockchain technologies or artificial intelligence (AI). These will be discussed in chapter 5.

URBAN CO-GOVERNANCE AS A DRIVER FOR NEIGHBORHOOD POOLING ECONOMIES

The previous examples illustrate that one critical aspect of our framework, urban co-governance, is the creation of public-private-science-social-community partnerships (5P) as a policy and legal tool to ground collaboration between the public administration and community-based social innovators through co-design processes geared toward projects that serve the public interest—both in the sense of serving public versus private values and in serving community needs (Iaione 2016; Foster and Iaione 2019). The construction of a non-authoritative (horizontal, collaborative, and polycentric) relationship between the local government and city inhabitants in which they are on an equal footing requires changes in the relationship between public and civic actors as well the private sector. The public administration must facilitate the creation of platforms for this collective action and be willing to invest resources on spearheading the establishment of collaborative ecosystems across the city. Private and civic actors must be able and willing to invest time and build skills and capacity to assist local communities, particularly the most marginal and vulnerable, to undertake stewardship of urban goods, services, and infrastructure.

Policy and legal tools such as pacts of collaboration or neighborhood agreements, as we have seen emerging from Italian cities, can be understood as a public institutional governance mechanism but also as a way to facilitate new pooling economies for the poorest and most disinvested communities. There is no single accepted definition of the many collaborative economic paradigms (Selloni 2017; Rinne 2020; Botsman and Rogers 2010), with many equating it with a *peer-to-peer* (Bauwens and Pantazis 2018), a *collaborative* or *sharing* economy (Rifkin 2014; Sundararajan 2016) and others with a *platform* economy (Montalban et al. 2019). We define, for our purposes, a pooling economy to be facilitated by multisectoral, multistakeholder platforms that produce goods and services with affected communities that can be stewarded or co-governed by those communities, along with other stakeholders. In a pooling economy, goods and services are not just shared but also co-used, co-managed, co-produced, and even collectively owned by the different users, which in the most complex cases include other stakeholders, such as public authorities and private operators.

The Turin case is an example of how the creation of an urban co-governance scheme can spur, support, and sustain collaborative economies geared toward the most vulnerable of urban residents. We ended chapter 3 by describing the recently adopted regulation in Turin that involves partnerships of established communities and different urban actors for regeneration, care, and management of shared urban assets. Here we describe briefly the ways that the urban co-governance approach embraced by that regulation addresses economic issues in an innovative manner.

Turin has historically played a key role in Italian industrialization, driven mainly by automobile manufacturing (FIAT cars were manufactured there). The historian Lewis Mumford (1938) compared Turin to cities like Pittsburg, Lyon, and Essen, known as heavy-industrial economies. Turin was a center for Italian domestic immigration from its eastern and southern regions, where migrants came to work in the car manufacturing industry, often living in poor conditions. In the 1980s, with the decline of the car manufacturing industry and the related job losses and political corruption scandals, the city entered an economic and social crisis. Starting in the 1990s, the city leadership began to focus its efforts on kickstarting an urban-planning-based economic development strategy (Galanti 2014, 160–161).

Thus, urban policy makers and private foundations have attempted to revitalize and diversify Turin's economy, including efforts to create a creative and cultural sector, enhanced tourism, private entrepreneurship in the ICT sector, and R & D specialization. The city was seriously impacted by the economic crisis in 2008, with the unemployment rate reaching a high among northern and central Italian cities. The city is still reeling today from the economic crisis (the current rate of unemployment in the city is 9.4 percent, higher than the average in northern and central Italian cities) and is coping with declining suburban neighborhoods on the periphery of the city, in part the result of de-industrialization processes, where vulnerable communities and migrants continue to live. Since the 1990s, Turin has coped with the challenge of stimulating its economic development and providing urban welfare and the regeneration processes of building and urban public spaces in the city outskirts. It has done so mostly through EU funds for urban regeneration. The implementation process of the regulation on the urban commons leveraged also the Turin Co-City and Co4Cities projects supported financially by the Regional

Development Fund of the European Union through its Urban Innovative Actions initiative (UIA) and the URBACT program, as well as several other initiatives aimed at stimulating social entrepreneurship and social or impact investing such as Torino Social Impact and Homes for All.

The underlying objective of all these initiatives is to transform abandoned structures and vacant land into hubs of inhabitants' participation in order to foster community-based multi-stakeholder cooperation and to create social enterprises that will contribute to the reduction of urban poverty. The regeneration of abandoned or underused spaces in different areas of the city is expected to contribute to the creation of new jobs in the social economy sector through the creation of new enterprises at the neighborhood level. These enterprises are expected to stem out of the close collaboration and working relationship between the local government, residents, and the social and technological vibrant innovation ecosystem.

A key role is played in many instances by the network of Neighborhood Houses (Case del Quartiere) in the city. These Neighborhood Houses (NHs) consist of eight neighborhood community centers located in formerly distressed or abandoned spaces that have been regenerated and made available to the local community for civic, cultural, and educational uses (Caponio and Donatiello 2017). The networking of these houses has been promoted and supported over the past ten years by the local government and other public and private stakeholders to foster cooperation and free exchanges between the houses and to innovate neighborhood level responses to the needs of vulnerable individuals, specific groups, and the community.

To facilitate the construction of collaborative economies, the Neighborhood Houses act as pooling economies development agencies and are a critical institutional tool enabled by the Turin regulation. Much like the social centers in Reggio Emilia, the Neighborhood Houses in Turin follow the quintuple helix approach that we have argued is part of the framework of urban collective governance. They bring together at least five possible urban actors or stakeholders in a collaborative institutional format to instigate peer-to-peer production of neighborhood goods and services. The eight Neighborhood Houses are highly differentiated institutions, realize through public (city, regional, and EU) and private funding (private and philanthropic foundations). The Neighborhood Houses network is a key actor, structuring the urban communities' capacity to access the opportunities offered by the Regulation. Through Neighborhood Houses, Turin's inhabitants can find all the information regarding the Turin's urban commons and innovation ecosystem and the opportunity that it offers. They can also access support for drafting proposals of pacts of collaboration as well as have the opportunity to meet other city inhabitants interested in establishing a cooperative or a foundation to take care of or regenerate the same urban commons.

One of the differences between the Turin Neighborhood Houses network and the Reggio Emilia Social Centers is the focus on regeneration of dismissed urban assets as the first goal of the partnership. The object of the intervention is always a city-owned building, dormant or dismissed (Bertello 2012). The neighborhood house Cascina Roccafranca, for instance, in 2007 was a privately owned farm of 2,500 square meters that the city purchased through EU urban regeneration funds. Then the city of Turin, together with local NGOs, established the Cascina Roccafranca Foundation for the recovery and management of the farm, which had been turned into a community center, and the promotion of self-organization among city inhabitants. At the Cascina Roccafranca, city inhabitants can access a variety of services such as legal counseling, potable water distribution, and an orientation service for migrants and vulnerable people on the hospitability services in Turin.

The city launched a call for proposals for potential pacts of collaboration. The call had a high response rate with approximately 115 proposals submitted in the first phase. The project then moved to a co-design phase, whereby a select group of proposals for collaboration pacts would undergo a process of fine tuning and aligning with the project's larger policy goal of poverty reduction as well as other local government goals, and a feasibility evaluation. After the first year of the project's implementation, the first pacts of collaboration were finalized and signed. The first pacts demonstrate the variety of ways that the Turin *civic deals* and pacts of collaboration are geared toward collaborative economies (Ferrero and Zanasi 2020). A brief overview of some examples of these pacts may help understand the kind of neighborhood-based pooling economies that urban co-governance mechanisms can jumpstart.

One of the pacts (the Casa Ozanam Community hub pact) is to transform a previously abandoned building into a neighborhood house in partnership with local community-based associations and cooperatives active at the neighborhood level to deliver social services and access to other critical infrastructure for the neighborhood. The target populations are teenagers and young adults, and the project will create an open community garden, offer sport facilities, and organize cultural events and workshops.

Another successful pact, Cucina del Borgo, is focused on food and culture through the establishment of an adjunct restaurant. The purpose of this initiative is to create recipes with inhabitants of the neighborhood, enabling them to recount their own stories. Recipes are accompanied by pictures and narratives of the residents, and the food is then served in the adjacent restaurant. The project helps to shape the identity of the community and to revitalize it by highlighting the richness of neighborhood culture and history.

The Habitat project and pact will offer childcare and shared spaces in which to organize activities to help alleviate economic distress and support job searches for working parents. These activities include networking meetings, training, and workshops. The pact is geared toward creating these spaces for support services and new forms of social service assistance within a building already hosting health services.

The Falklab project has been designed to renovate an underused building within a school complex and use it to offer artistic workshops for teenagers. Learning laboratories and networking events are designed to connect and bridge diverse communities and actors in a thick and complex urban environment through the creation of essential social infrastructure.

These examples illustrate that the Turin ecosystem recognizes that the collaborative economy can develop at the neighborhood level through institutional arrangements representing neighborhood, local, and public actors. Community residents can work together in a deliberative and collaborative manner with other stakeholders, at a defined geographic scale, to maintain or regenerate the assets and services within their neighborhood or district. These assets are then transformed into collaboratively managed resources, goods, and services.

Other cities are also investing in similar approaches geared toward collective self-entrepreneurship of entire neighborhoods. The city of Reggio Emilia, for example, through its collaboratory specifically experiments on hybrid business models capable of jump-starting neighborhood economies. An example of the kind of project that emerges from the collaboratory is the Coviolo Wireless project, a community wireless mesh network (CWN). This project recognizes that technological and digital infrastructure can be the object/goal of urban co-governance, similarly to the New York City example. CWNs are self-constructed networks that provide solutions for access to the internet in vulnerable rural or urban communities (Oliver, Zuidweg, and Batikas 2010). The Coviolo Wireless project in Reggio Emilia was formed through a citizenship agreement from a partnership between the regional digital network's infrastructure operator (Lepida) and a neighborhood community center managed by local residents. The regional company operating the digital network infrastructure was convinced by the city to extend the main network infrastructure to allow interconnection with the neighborhood wireless infrastructure. The neighborhood community center now owns the neighborhood wireless network infrastructure and serves as an intermediary between the users and the main network infrastructure, managing network access and providing affordable broadband services (between €13 and €16 monthly) to neighborhood inhabitants. The infrastructure is economically self-sustained through users' fees.

FINANCING URBAN CO-GOVERNANCE

Finally, we turn briefly to the existence of two primary ways to finance the commons. There are many potential modes of financing, which range from crowdfunding to fundraising. In our view, there are two most promising approaches on the horizon. The first builds on project finance techniques that can be redesigned to serve local communities' goals. The second is intended to piggyback on a new trend in corporate finance—the *environmental, social, and governance* (ESG) approach that lies at the heart of the global initiatives currently taking place to give birth to what has been called *green and sustainable finance*.

From a project finance point of view, an Urban Commons Foundation is one of the most advanced financial instruments for the governance of urban commons. This Foundation is a feature of the new Turin Regulation described in the chapter 3 and in this chapter. It is also the first financing tool introduced by an Italian local policy. To establish a commons foundation, a city transfers one or more or assets to a foundation established for the sole purpose of managing shared urban resources in the general, public, or common interest. The transferred assets constitute a patrimony of the Foundation and thus are restricted from being sold or transferred. The basic idea is to endow a new legal entity, the Urban Commons Foundation, the rights of stewardship and ownership over critical assets or properties for a particular community. The Foundation has the duty to carry out the kinds of interventions and activities that would normally be carried out by the community through recognized forms of co-governance such as pacts of collaboration, neighborhood agreements and civic and collective urban use rights. The establishment of such a Foundation aims to maximize the social and economic value of these assets for future generations, which is particularly relevant for shared, common resources with importance to heritage. The local government, as such, divests itself of the asset in the service of a collective governance arrangement, and each decision regarding the management and use of the assets is the responsibility of the Foundation, with the general and common interest as a guiding star.

The creation of this kind of Foundation occurs in two stages. In the first, only the rights of use of the assets are endowed to the Foundation for a limited time. After this trial period, the asset can be permanently transferred to the Foundation. Given the importance of the decision to relinquish a municipally owned good, the establishment of the Foundation must be approved by the city council. The constitution and the status of the Foundation are drafted by a group composed of members of the administration, democratically designated representatives of the community of reference, and experts chosen from the Register of Guarantors.

The New Turin regulation also includes some rules, principles, and criteria such as the democratic composition of the control and decision-making bodies to guarantee the representation of all the actors involved in the governance of the urban commons. The expectation is that the assets would be given back to the local government to maintain its public use should the Foundation ever be dismantled. Finally, the city can join the Foundation, but only using its private law capacity/autonomy and not bearing its administrative powers.

The Foundation and related financing represent an evolution from the Naples and Bologna experiences described in chapter 3. Contained in article 24 of the New Turin Regulation, "Self-Funding," the main innovation represented by the regulation is the possibility of public sponsorship and supported profit-oriented economic activity as a form of self-financing for the care, regeneration, and co-management of urban commons.

Alongside the Foundation, the New Turin Regulation introduces other fundraising and urban commons finance tools. Similarly to the New York City example described previously in this chapter, the Turin case pays attention to equipping communities with the necessary skills and tools to be able to open dialogue with financial institutions and to generate long-term investors' interest. There are several ways in which the regulation helps to sustain pooling economies benefiting local communities as they steward shared urban resources. For instance, the regulation creates tax and fee exemptions, and it assumes the cost of utilities or other services to support co-governance and management of shared resources. As an example, a contract to engage in services or to purchase goods to support the urban commons might be exempted from municipal taxes and fees, on the grounds that the activity is for the use of public land. As another example, local government might allow city staff to work on the projects or to engage in related projects as a form of community service. The local government can also directly carry out activities or provide essential goods for the project. Although the local administration cannot allocate direct contributions or subsidies to these projects, the projects can participate in public calls and tenders, and the city can facilitate fundraising for the project partnership. The public administration can also take on the cost of utilities, depending on the methods and limits defined in the provision approving the partnership. With regard to financial support, the administration cannot allocate direct contributions or subsidies to civic subjects, but the latter can participate in public calls and tenders and the city can facilitate fundraising for the civic deal partnership.

Risk prevention within the civic deal is guided by a principle of civic autonomy and mutual trust. Civic subjects are not workers but rather subjects with rights comparable to those granted to volunteers. The civic deal might be completed with documents describing the places and risks specific to the assets and activities carried and the measures to be adequate. Civic subjects are considered custodians for the urban commons, but the partnership regulates the punctual division of responsibilities with the administration.

A second financing innovation for the commons is the sustainable finance framework that involves a rethinking of public and private investment metrics. These metrics or indicators can measure, for instance, the degree of multistakeholder cooperation and pooling economies as forms of collective neighborhood-based business models supporting the efforts of city residents and urban stakeholders to partner and cooperate toward shared goals. Purpose-driven private investors (Johnstone-Louis et al. 2020) or patient, long-term institutional investors (Prodi-Sautter 2018) could potentially be interested in stimulating economic growth and can generate societal impact. This model of blended profit and socially motivated investments has been adopted in the US, the UK, and the EU (Bugg-Levine and Emerson 2011). Measuring the impact produced by the civic deal is of the utmost importance under the New Turin Regulation. The modalities for conducting monitoring and evaluation activities are agreed upon as part of the civic deal. The documents produced for this purpose must be comparable, accessible, verifiable, complete, and produced at least yearly by an independent evaluator. The evaluation reflects the social and economic impact of the activities carried out within the civic deal and must be widely disseminated.

A related sustainable finance tool, as we underlined in our previous work (Foster and Iaione 2019) is the Community Interest Company (CIC) (CIC Regulator n.d.). Focused mainly on providing benefits for the local communities and brought into existence through statutory clauses that ensure that the assets are used for the benefit to them (Cho 2016), the CIC can be set up as a social enterprise whose profits are reinvested exclusively for its social purposes. In the UK legal framework, CICs can include local community enterprises, social firms, cooperatives, or national or international organizations (UK Government 2017). A similar corporate model is embodied by the US model of the benefit corporation, which is a business entity that pursues the general interest as a statutory requirement. It is a popular corporate form of social enterprise (Murray 2012). Cooperative alternatives to the corporate model are the Italian *cooperatives* and specifically the *community cooperative* model (Mori 2014; Grignani et al. 2021). One example of CIC operating at the urban level is Hackney Community Transport (https://hctgroup.org/). Founded in 1982 by local communities in the London borough of Hackney to provide affordable transportation to facilitate NGOs and other social actors in their support to vulnerable residents, HCT evolved over time, and eventually Transport for London awarded the group the contract to operate a London Bus Route from the

Islington to Homertown Hospital. HCT grew and expanded into other boroughs and then outside London, eventually merging with other organizations offering a similar service. It became a CIC in 2007 (Nicholls 2009).

The analysis in this chapter demonstrates how much work and practice has been carried out but also how much needs to be developed to realize more robust institutional tools to support the vision of urban co-governance laid out here. As we have argued, what is important for our model is not simply community engagement or inclusion but rather educational, digital, and financial tools to build and enable the capacity of local communities to be fully engaged and empowered. Another real challenge is the scaling up of these experiences. Financing programs designed to support the kind of projects mentioned in this chapter, supported by EU funding programs such as UIA, URBACT, and H2020, showcase the possibility of providing these tools. However, the private sector role is still evolving and includes the possibility of supporting urban co-governance through social finance, impact finance, and green and sustainable finance initiatives. These are at present unrealized options to guarantee the self-financing and self-sustainability of the urban commons in the future.

But when speaking of new financing mechanisms, the future seems to be brighter for communities willing to claim their co-governance rights on the urban commons. The Social Taxonomy Report (Platform on Sustainable Finance 2022) introduces an objective very coherent with the co-governance approach: *(ii) making basic economic infrastructure available to certain target groups. This objective focuses on people in their role as members of communities*. It seems to reinforce a policy commitment to co-governance when it comes to some of the sub-objectives, which foresees for communities the recognition of *land rights* and a call on *improving/maintaining the accessibility and availability of basic economic infrastructure and services like clean electricity and water*.

Finally, the co-city approach finds further support in the new EU Cohesion Policy framework, which allows the informal cooperation of citizens through European Groups of Territorial Cooperation and through the new European Urban Initiative to develop further community-based integrated approaches to urban development.

5

THE CO-CITY DESIGN PRINCIPLES

The book to this point has provided the conceptual and experiential foundation for a new urban governance model that we call the *co-city approach*. This framework imagines the city as an infrastructure in which a variety of urban actors cooperate and collaborate to govern and steward built, environmental, cultural, and digital goods through contractual or institutionalized a particular public-community partnerships (PCPs) and public-community-private partnerships (PCPPs): the *public-private-science-social-community partnerships* or simply 5P. These partnerships involve cooperation and collaboration between civic, social, knowledge, public, and private actors that support the creation and governance of shared and common resources by an identified group of people, a community, vested with the responsibility of maintaining and keeping accessible (or affordable) the resource for future users and generations. These common resources occupy a middle ground between public and private goods and between the state and the market. They represent new and innovative forms of urban goods and services geared toward supporting the most disadvantaged, marginal populations and communities. A co-city is based also on the idea of polycentric governance, which allows for the co-production and co-governance of a variety of shared resources in multiple but mutually supportive institutional arrangements throughout a city.

In this final chapter before the conclusion, we offer a set of design principles extracted from the concepts described in the previous chapters, as well as our empirically driven co-cities research project. The co-cities project canvassed over two hundred cities and over five hundred projects and policies within these cities and more closely analyzed case studies of community or city-level initiatives that represent new frontiers of cooperative or collaborative urban governance, inclusive and sustainable local economies, and social innovation in the provision of local goods and services. The data set includes examples of projects and public policies from different types of cities, including some of the groundbreaking policy experiments that we describe in this book. Both the community-led examples and those institutionalized in the local government are important data points and empirical inputs into the larger effort to explicate the dynamic process (or transition) toward a city in which urban commons are present, to one in which urban commons emerge and are supported and enabled by the state.

All the case studies are published on the web platform Commoning.city. For the purpose of this book, we extracted cases that were more robust (at least three variables out of five were assessed moderate to strong) or that were outliers (presented extraordinary features compared to all cases in one of the variables), and we analyzed, through qualitative interviews in addition to desk research, 140 cities with 283 cases within them. The appendix contains a summary of the data we collected and analyzed.

Our co-cities framework is defined by five principles: (1) co-governance; (2) enabling state; (3) pooling economies; (4) urban experimentalism; and (5) tech justice. Our research has shown that these recurring characteristics, methodologies, and techniques best define the ways in which the city can operate as a cooperative space in which various forms of urban commons can emerge and can be economically, socially, and ecologically sustainable. Some of the design principles described in this chapter will resonate with some of Elinor Ostrom's design principles, as we indicated in chapter 2, and others reflect the reality of constructing common resources in the context of contemporary urban environments. In the concluding chapter, we reflect briefly on the challenges we continue to face in the application of the co-city design principles and pathways for future study and research.

PRINCIPAL 1: CO-GOVERNANCE

The co-city approach rests in part on the body of theory developed by Elinor Ostrom and others, encompassing a range of approaches to shared, collaborative, and polycentric urban governance mechanisms that structure cooperative action between and among different types of urban actors. Scholars have referenced this type of multi-actor governance by different names and definitions: these include collective governance (Ostrom 1990), self-governance (Ostrom 1990; Harvey 2012; Nielsen 2015), shared governance (Laerhoven and Barnes 2019), collaborative governance (Freeman 1997; Ansell and Gash 2008; Bingham 2009 and 2010), cooperative governance (Wilson et al. 2003), co-governance (Kooiman 2003), and polycentrism (Ostrom et al. 1961; Ostrom 2010b). We have drawn from this research over the years in reflecting on the kind of co-governance that we have observed and applied in our own work within cities, as well as the cases and examples that we surveyed as part of our co-cities project. We also recognize and build upon the evolution of urban governance approaches, such as participatory budgeting, in which decisions are made cooperatively between local officials, residents, and community-based or civic organizations. These approaches represent an important move beyond resident participation and consultation in local decision-making processes.

Our approach to co-governance also builds on the extensive literature defining various combinations of open, productive, knowledge, constructed, and infrastructure commons and the peer-to-peer production mechanisms associated with them. These definitions have been developed by Carol Rose (1986); Yochai Benkler (2016); Elinor Ostrom and Charlotte Hess (2007); Michael Madison, Katherine Strandburg, and Brett Frishmann (2010, 2014, 2016). Finally, Tine De Moor (2012) helpfully suggests that we think about the commons as consisting of three dimensions: a resource system, a collective property regime, and interactions between the resource and its users. All three come together to create a common pool or commons institution. Each of these articulations and iterations of cooperative governance have the potential to foster democratic legitimacy, transparency, and social inclusion (Bang 2010).

In our view, co-governance embraces and entails the collective management and ownership of urban assets that provide resources and critical services for the well-being of the most vulnerable urban residents. Some of the most forward-thinking examples of city policies and practices, analyzed in chapters 3 and 4, adopt a concept of co-governance that entails not only a relationship between the public authority and the social or civic sector but also various combinations of relationships between those actors and private actors in the pursuit of the common good and general interest (Kooiman 1993). As such, we recognize that co-governance practices can evolve, as reflected in the evolution of public policies described in chapter 3, to enable and recognize the development of urban commons throughout a city. The earliest policies, represented by cities like Bologna, Naples, Barcelona, and Madrid, enabled shared governance between urban actors and local officials instituted through contractual collaboration agreements. As these approaches to enabling and facilitating urban commons evolved, we began to see the emergence of more complex multistakeholder partnerships, and then more independent but networked co-governance structures that were forged by place-specific actors and established in new neighborhood institutions.

Much like Arnstein's ladder of citizen participation (Arnstein 1969), we conceive of urban co-governance as a ladder involving the evolution of steps of co-governance: shared, collaborative, and polycentric governance. This co-governance ladder can be used to construct urban commons public policies and projects in a specific local context by nudging those involved toward a higher gradient in a cooperative institutional ecosystem. As co-governance proceeds along this ladder, the multistakeholder governance arrangement adopted by the actors involved moves from shared governance to collaborative and potentially polycentric governance. We thus define co-governance as a multistakeholder governance scheme whereby the community emerges as a key actor and partners with at least one of the other four actors or sectors in our *quintuple helix* governance scheme—the public sector, the civic sector, the private sector, and the knowledge sector. In addition to collaborating with other actors, co-governance entails the interactions between actors that constitute a specific governance arrangement operating within a polycentric network of agreements, institutions, or informal arrangements throughout a city.

In sum, there are different levels or steps of co-governance: *shared, collaborative*, and *polycentric* governance. This urban co-governance ladder, so to speak, is a typology of increasingly robust levels of co-governance corresponding to the extent of diversity among the actors, distribution of power between them, and responsibilities and benefits within the partnership.

SHARED GOVERNANCE

Shared governance encompasses bilateral interactions or partnerships between, for instance, public authorities and urban residents or communities. Shared governance exists, we observe, in the management or stewardship of small-scale resources such as community gardens or a neighborhood park or square. These bilateral partnerships or agreements are typically between local government authorities and residents who volunteer to take care of, regenerate, or manage a single urban resource in order to improve the quality of urban spaces for users. Improving the quality of the resource could also involve producing essential goods and services, much in the way that urban green spaces provide food and recreational amenities in neighborhoods lacking sufficient levels of them.

Shared governance is akin to Elinor Ostrom's notion of self-governance, a crucial feature of collective action to manage common pool resources (Ostrom 1990). Self-governance more broadly refers to the "situation in which actors take care of themselves, outside the purview of government" (Kooiman and Bavink 2005, 21). It is a prerequisite for more complex and dynamic forms of co-governance. Self-governance might seem to some to be just another form of privatization or even deregulation of public goods and services; however, we view it more as a form of *re-regulation*, meaning that control and management remains in the realm of the *public*—not the central government but rather the community of residents that use and depend on the resource (Kooiman and van Vliet 2000, 360).

COLLABORATIVE GOVERNANCE

Beyond bilateral public-community partnerships, there are more expansive relationships and interdependence among many different types of stakeholders, including public, community, private, civic, and others, which come together to construct, manage, and govern common-pooled resources. Typically, these partnerships consist of a minimum of three of the actors of the quintuple helix that become deeply engaged over time in constructing and supporting institutional arrangements to support resource stewardship. These multistakeholder arrangements can be informal in nature or more formal and institutionalized but are the product of a process of deep multistakeholder engagements and interactions (Kooiman 2003, 97).

Collaborative governance represents the evolution from agreements or pacts that foster collaboration among stakeholders to governance structures or legal entities that are cooperatively owned by or linked to the actors of the quintuple helix. Collaborative governance arrangements are often realized, as we explored in chapter 3, through the implementation of public policies that enable nongovernmental stakeholders to manage public resources collaboratively (Ansell and Gash 2008). These policies often promote multisector cooperation and partnerships between profit and not-for-profit actors, relying either upon existing or new relationships and social networks (Cepiku, Ferrari, and Greco 2006).

The shift from shared governance to collaborative governance also tracks the move from the management of a single small-scale urban resource to the management of larger-scale resources at the neighborhood or the city level. One main manifestation of this collaborative governance is the embrace of public utilities that produce local public services (Non-Profit Utilities [NPUs]) and involve many stakeholders in collective property ownership or management of the public service management, forbidding the distribution of dividends to its members. Public utilities such as the water distribution company in Wales, Glas Cymru, can be institutionally engineered or redesigned as user/community/worker-owned and -led cooperatives in which networks or infrastructure are run cooperatively. Community co-ops, community land trusts, urban civic uses, and other cooperative legal tools mentioned in chapters 2 and 3 are also examples of this type of collaborative governance of large-scale resources and are excellent examples of what Ackerman (2004) refers to as co-governance cases.

POLYCENTRIC GOVERNANCE

Cooperative governance arrangements, whether driven by informal social norms or formal institutions, eventually can evolve into a polycentric system on the city level. The polycentric approach to governance was first proposed by Vincent Ostrom, Charles Tiebout, and Robert Warren to connote "many centers of decision-making which are formally independent of each other" but which "may function in a coherent manner with consistent and predictable patterns of interacting behavior" (Ostrom et al. 1961, 831). In polycentric governance, actors and cooperative institutions are autonomous centers of decision making as they interact and learn from each other while maintaining their respective responsibilities.

The policies analyzed in this book, especially in advanced cases such as Turin, Reggio Emilia, and Barcelona in Europe or the cases of networks of community land trusts within US cities, represent the forms of shared governance and the forms of self-governance that can generate polycentrism. In other words, these examples reflect the creation of a multiplicity of independent decision-making arrangements or institutions (Fung 2001, 87) consisting of a plurality of urban actors (the community, local businesses, knowledge institutions, and civil society organizations) that are managing, together with public institutions, shared urban resources. This is the highest, most advanced form of co-governance in our co-city approach (Iaione and Cannavò 2015).

As we briefly explained in chapter 1, Elinor Ostrom explored a polycentric approach to governance in the context of determining the efficiency of an array of public and private agencies or actors engaged in providing and producing public services in metropolitan areas. Her study, which focused on the provision of law enforcement services, found that residents of small communities were more satisfied with their locally organized police forces than demographically similar communities who were served by a centralized police force (Ostrom et al. 1973). Metropolitan areas served by multiple jurisdictions and producers of public goods, she found, benefit by having more choice in the provision of public goods, are more likely to utilize innovative approaches to public goods provisions (including citizen co-production), and can learn from each other's performance in providing those goods (Ostrom 1990).

Polycentric governance systems can, and do, apply to the provision of local goods as well as to global challenges such as climate change. Instead of a global, top-down regime in which lower levels of actors carry out the mandates from above, a polycentric approach "provides for greater experimentation, learning, and cross-influence among different levels and units" of decision making (Cole 2011, 395). These governing units can include a myriad of nongovernmental organizations, local neighborhood associations, and individual property owners who can play an important role in governing resources (Cole 2011, 397).

A polycentric approach to local governance can be a metadesign principle for recognizing the city as a commons. Collective governance of shared urban resources constitutes autonomous centers of decision making, whether they are community gardens, neighborhood parks, or community land trusts. When facilitated and supported by state actors, recognition of these collective efforts can be configured as a system, supported by local law and allowing for the coexistence of multiple centers of governance with different rules, values, perspectives, and interests (Aligica and Tarko 2012, 245–260). Another crucial feature of polycentrism is its capacity to enhance learning through experimentation, or the trial-and-error approach, allowing the system to change, adapt, and self-correct (E. Ostrom 1998; Carlisle and Gruby 2017, 7). Polycentric systems are therefore dynamic governance arrangements and not static, even when institutionalized.

PRINCIPLE 2: ENABLING STATE

Our next principle, enabling state, focuses on the role of local public authorities. We have explained much of the conceptual and theoretical underpinnings of this principle in chapters 2 and 4. As we have indicated, the presence of a governmental unit or a policy that facilitates and enables co-governance of urban infrastructure and resources is a key factor for the success of urban commons, as scholars have previously noted (Foster 2011; Nagendra and Ostrom 2014). As we explained in chapter 2, one of Elinor Ostrom's design principles for the long-term sustainability of the commons was that a community's right to collectively govern a common pool resource and to devise its own rules is recognized and respected by outside central authorities. Such recognition renders the rules easier to monitor and enforce, according to Ostrom. The role of central authorities or the state is even more present in the creation and sustainability of the urban commons, given that the local government typically retains regulatory control and, in some cases, proprietary ownership of these resources. The policies discussed in chapter 3 illustrate the many ways that municipal authorities have been instrumental in the recognition and constitutive creation of urban commons.

There are other, more normative reasons for the importance of an enabling state. The core one, for us, is that the enabling state principle embraces a *resurgence* of the state after a decline, beginning from the late 1970s (Cassese 2017). The rise of neoliberalism, marked by the retreat of the state and the prevalence of the private sector in providing for basic goods and services, has been accompanied by a rise in stark economic inequality. The enabling-state principle reflects the view that the state can and should play a catalytic role in directing change by helping to form new institutional structures, transform landscapes, and create or shape new economies (Mazzucato 2015, 3). As cities have become increasingly more complex to govern, we have seen the rise of more networked forms of urban governance through the expansion and diversity of actors involved in official decision making and policy making processes, as well as more varied institutional arrangements (da Cruz et al. 2018). The state's role in this time of transition often becomes that of a manager of different centers of power and subsystems, helping to network and create their interdependence (Jessop 2016, 248). This dynamic is particularly obvious in the cities that we highlighted in chapters 3 and 4, in which local officials are facilitating robust neighborhood-level institutional arrangements and connecting them through a supported polycentric system of urban commons.

The conceptualization of the enabling state that we are offering here and that emerges from the previous chapter revolves around an open and collaborative governance methodology. It embraces public-community and multistakeholder partnerships without giving up the state's regulatory power over shared resources. This principle supports the urban commons by investing in collective community efforts, transferring resources to support those efforts (such as available land and funding), and providing technical support to increase the capacity of the actors involved. The enabling state adopts an experimental approach to policy making that bring together residents, NGOs, civic organizations, knowledge institutions, social innovators, and businesses to co-design and co-create local public policies for environmental, economic, social, and cultural progress. When the governmental unit has permanent or long-term staff members acting as the service designers of co-design processes, these designers host urban experiments in urban living labs where co-design and co-working sessions can take place (Franz 2015; Steen and van Bueren 2017). An enabling government is in fact a government that uses services or policy co-design as a tool for decision making and planning, as well as a tool for the management of essential services and infrastructure or assets.

Another way to think about the principle of the enabling state is to think of the state as a platform, in which the state entity does not seek to guide the co-creation process itself but rather to play its role from a distance, taking a bottom-up approach and a supporting role in a network of relationships (Iaione 2018). The platform state disrupts the monopoly that the central government has over deciding what is in the public or general interest and instead becomes comfortable with a network of actors making these decisions for their communities and the resources within them. As such, whereas the state can be an indispensable actor in facilitating and sustaining new collective action in communities throughout a city, our research has demonstrated the need for active participation of other actors to support these efforts. Academic institutions and social investors are critical players in scaling and networking community-based efforts that are key to creating local, adaptive forms of co-governance throughout a city.

A local administration willing to become an enabling platform will fund and invest in urban co-governance and multistakeholder partnerships with knowledge and investment partners that can be the trigger for new circular, tech-based, and community-owned economies. Strategic use of public procurement, aimed at creating jobs to confront digital transformation and ecological transition processes, new community enterprises, and social businesses in deprived neighborhoods are another new and innovative way to fund collective institutions (Iaione 2018). The research hypothesis embedded in this principle—the enabling state—is rooted in the relationship between urban co-governance and economic democracy. It understands the city as a facilitator of significant investment in communities not only as spaces for civic engagement but as productive units of inclusive and sustainable economic development.

We must account for something that Robert Dahl (1967) suggested, namely, that the city is the best unit of measure for democracy: small enough for participation and big enough for the contribution and influence of individuals to be significant. The sublocal level appears to be even more important, if we consider that in most of the cases that we analyzed in chapters 3 and 4, the pilot projects occur at the neighborhood level. We argue that cities cannot create urban commons without significant action and collective efforts occurring at the neighborhood level. Our analysis of cases and our own experience working in cities on experimental approaches to the urban commons indicate the need to build a strong connection between urban co-governance and economic democracy at the neighborhood level.

Our evaluation of the Bologna regulation's initial implementation, for instance, suggests that Bologna may have initially underestimated the need to create multistakeholder partnerships promoting the systematic establishment throughout the city of neighborhood-level collective economic institutions aimed at jump-starting sustainable and inclusive economic development in distressed areas of the city (Iaione and De Nictolis 2021). These areas suffer from chronic underinvestment and access to basic services and infrastructure and would greatly benefit from self-sustaining urban commons institutions. One of the lessons learned from the Bologna case is the importance of channeling public and private support toward neighborhood-level collective economic units and granting collective rights of access, use, management, and ownership to *social purpose vehicles* collectively incorporated, controlled, managed, and owned also by city inhabitants, as in the Pilastro neighborhood project. This entails the need for a city government to invest in a policy strategy that targets neighborhoods, not only as spaces for civic engagement but as productive units of inclusive collective economic development. This also involves conceiving of and treating the city as a polycentric entity in which neighborhoods act as decentralized engines of inclusive collective economic development through which communities can identify common interests and begin to co-produce or co-manage services with centralized coordination by the city government. This is where the concept of pooling economies becomes very relevant.

PRINCIPLE 3: POOLING ECONOMIES

As referenced in chapter 2, we have observed that many kinds of urban commons exist as a product of what we call *social and economic pooling*.

It is important to note that our use of the term *pooling* is not a reference to the features of a common pool resource as Ostrom and others use that term (Ostrom 1990). Our use of the term pooling instead describes the process of different sectors or actors combining their efforts to share resources, collaborate, and cooperate to create and steward urban goods, services, and infrastructure. The pooling of capacities and resources thus makes possible the co-production and co-creation of collectively owned or collectively managed economic ventures, creating equal opportunities for the community as a whole and not solely for the individual (Rawls 1971; Sen 1992). These *urban pools* can generate new collaborative, circular, and solidarity economies at the neighborhood, district, and city level. Resources become an urban *commons* or part of a common *pool* through these collaborative practices and ventures aimed at sharing existing urban resources, generating new resources, producing new public services, and coordinating urban networks across the city.

Social and economic pooling is, in other words, the signal of a transition from an urban co-governance scheme in which different neighborhood actors share, co-manage, and regenerate the urban commons toward an urban co-governance scheme based on urban pools in which the same actors coalesce to transform neighborhoods into social and economic enabling platforms. Anna di Robilant has noted that common-ownership regimes (e.g., land trusts, limited equity housing cooperatives, neighborhood-managed parks, and community gardens) are those able to "make resources that are crucial to individuals' autonomy available on a more equitable basis" and foster new forms of autonomy (di Robilant 2012, 268–269). Pooling economies are connected to this "equitable autonomy" in that they enable those involved to enhance their capacity to debate ideas with others, collectively make decisions, and provide access to critical goods and services. Pooling economies, however, are not meant to substitute for the state nor for private economic actors in the production of goods and services. Rather, pooling economies utilize state resources and private economic actors to expand the capacity of neighborhood and community actors into collaborative and cooperative enterprises.

Pooling economics is rooted in a Polanyian understanding of the economy that facilitates a shift from productivity and competition as the basis for economic exchange to an economy based on social relations and reciprocity (Polanyi, 1944). This approach foresees a more interventionist role for the local authority, with increasingly risk-taking and proactive actions that support new local economies (Schragger 2016; Thompson 2015; Thompson 2019). In other words, pooling requires embracing a model of economic growth that has collaboration and reciprocity at its core (Mendell et al 2010; Mendell and Alain 2015). These forms of pooling economies are based on different degrees of sharing, collaboration, and cooperation between users.

The first and most well-known types of pooling economies have been digital. Yochai Benkler, as an example, has described commons-based peer production (CBPP) as a new system of production that emerges in the digitized economy, based on collaboration between peers and large groups of individuals, wherein the ownership is itself distributed (Benkler 2004). An example of a commons-based peer-produced resource is the online encyclopedia, Wikipedia. Another example, which we have referenced in this book, is the rise of community-based wireless networks (Tréguer and De Filippi 2015). The rise of *platform cooperativism* (Scholtz 2014), in which users/workers manage and own—with an organization that is inspired by the cooperative movement—their platforms to offer professional services and labor force, is an emerging example of CBPP. Platform cooperativism would allow workers to escape the often-unfair working conditions and economic treatment that sharing economy platforms offer and to keep revenues inside the group of users/workers or in the territory that the platform targets rather than distributing it to shareholders. Examples of this kind of platform include Coopify and Member's Media, Ltd. Cooperative. Coopify connects low-income workers in the sharing economy, such as movers or home care workers, who form worker cooperatives to engage a broader base of consumers and to scale beyond their current capacity. Member's Media, Ltd. Cooperative is an online platform, majority-owned by its users, which offers development and production support to aspiring microbudget filmmakers from diverse communities (Scholtz and Schneider 2016).

Similar to these *common-based peer production* models, our empirical results reveal many examples of the ways that urban pooling creates new kinds of shared, common goods in the housing, food, digital, energy, and cultural arenas in cities all over the world. Community gardens, wireless networks, co-housing, and land trust arrangements are most often the result of pooling together human capital, social networks, and existing urban infrastructure or public resources in efforts to create a structure of shared urban resources. Pooling also involves new fundraising and project finance tools that support collective efforts and ensure medium- or long-term sustainability of urban commons. Many of the examples that we have discussed previously illustrate the use of pooling economies to support the creation and sustenance of community land trusts in Boston or San Juan, the possibility to carry out for-profit activities inside renovated buildings through a pact of collaboration or a Commons Foundation in Turin, Italy, or a collectively owned and operated wireless community network in Reggio Emilia, Italy, as well as the impact evaluation of community co-management of city-owned buildings in Barcelona.

Our hypothesis is that urban commons are generating practices of social and economic pooling that can eventually be scaled up. Once the different actors involved in the co-governance of these resources understand the economic value and potential of joint action, these initially local innovations can be applied at a larger scale. Urban social and economic pooling is the bedrock on which co-governance partnerships should be designed and fostered to agglomerate social, economic, and institutional forces at the neighborhood level. Such alliances give birth to collective economic ventures that can produce job opportunities and that provide goods and services that benefit the communities where they are created. Social and economic pooling therefore is deeply connected to the distributive and social justice concerns that permeate the co-city approach.

The idea of an economy that is based on pooling and collaboration is very different from the *sharing economy* that involves a profit motive and is represented by gig economy platforms. This *crowd-based capitalism* is arguably replacing centralized institutions with peer-to-peer exchanges and mediated decentralized networks in ways that are disruptive to traditional market actors but not necessarily to property relations (Sundararajan 2016). It also generates externalities at the local level, prompting calls for centralized authorities to discipline the market through local regulation (Davidson and Infranca 2016).

In contrast, pooling economies foster a peer-to-peer approach that involves users in co-design and co-production and transform users into producers or owners of the delivery of goods and services. Pooling economies are capable of creating initiatives and platforms that are (1) collectively owned or managed; (2) multi-actor and cross-sectorial; (3) autonomous from but interdependent with other urban stakeholders; (4) aimed at generating a transfer of resources from the private sector or public sector to communities; (5) aimed at realizing the goals of the right to the city (e.g., right to housing and to universal access to public services and infrastructure such as broadband, energy, mobility, water, etc.); (6) sustainable, circular, climate-neutral and environmentally friendly; and (7) based on collective action at the neighborhood level (Committee of the Regions 2015).

PRINCIPLE 4: URBAN EXPERIMENTALISM

Our next design principle, urban experimentalism, represents an adaptive, place-based, and interactive approach to the design of legal and policy innovations that enable the urban commons. As commons public intellectuals and activists Silke Helfrich and David Bollier have noted, it is a mistake to equate commons with jointly "managed" resources only. This focus on the institutional characteristic of commons misses an essential part of the formula. Instead, commons are to be understood as "an organic fabric of social structures and processes" which involve the idea of "commoning" (Bollier and Helfrich 2015). Other scholars of the commons similarly underscore that the study and understanding of commons-based institutional arrangements are a product of applied, experimental, and local efforts by those involved (Poteete et al. 2010).

Urban experimentalism is an essential part of the process of constructing commons, including the institutional design of urban co-governance prototypes. We embrace the idea of experimentalism that represents a pluralistic, evidence-based approach to norm creation and policy making (Ouellette 2015). Experimentalism allows localized knowledge and diverse observation data to enrich the process for developing local policies and practices that support the urban commons. A key lesson from policy experimentation in general is the relevance of constant monitoring, learning, and adjustment as well as the capacity of policy makers to learn from failure (Dutz et al. 2014). The urban experimentalism that we advocate for, on the basis of our own experience working in cities on various policies and practices, contains three distinctive features: (1) an evaluative methodology that is data driven; (2) an experimental process that is adaptable; and (3) a process that is interactive.

As to the first feature, it is important to be able to evaluate the kinds of programs and policies described throughout this book using qualitative measures undertaken through surveys of participants, and quantitative measures where possible. Academic literature on the urban commons tends indeed to be normative, either heavily theoretical or explicitly ideological, and lacking an empirical focus. There are exceptions to this rule, as we discuss in chapter 2, but in the main, scholars writing about the urban commons are devoted to understanding the processes that result in collective action or cooperation in the governance of shared urban resources. There is much less focus on placing under empirical investigation, using established empirical methodologies, the many applied projects, and the policies that we have identified and mapped. For this reason, our LabGov team undertook an empirical evaluation of the most widely celebrated urban commons regulatory policy, the Bologna Regulation discussed in chapter 3. The legislation was the first of its kind to mention the urban commons as a subject of legislation and has since been copied or mirrored by many other cities in their policies, resulting to date in more than two hundred Italian cities that have adopted the Bologna Regulation in one form or another.

As we reported in chapter 3, the outcome of our analysis on the Bologna Regulation was to document the constellation and diversity of pact signatories, the kind of resources toward which they dedicated their efforts, whether pact signatories had a history of working together, whether the pacts increased social capital among participants, whether pacts were explicitly aimed at reducing social and economic inequalities in underserved areas of the city, and whether pacts were aimed at promoting local collaborative economies.

We gathered insights on the ability of a legal regulation to accomplish the broad goals that it embraced to institutionalize collaboration between government and city residents and to promote new forms of use, management, and ownership of urban critical resources to chart a new path toward social inclusion and justice. The results of the analysis also hold lessons for other local governments interested in adopting similar regulations and should serve as a cautionary tale in doing so without significant reforms. In fact, we further underscored the importance of putting in place a process for arriving at the right legal tools and policies as well as implementation that is adaptable to local conditions.

The second feature of our urban experimentalism principle is *adaptability*, which means putting in place structures and processes able to explore the right policies and mechanisms with communities and other local partners. Urban experiments are place based or place specific and are put in place taking into consideration the stakeholder network and the historical and cultural variables of the context, akin to experimental scientific research labs (Karvonen and van Heur 2014). Experiments organized within such cities as laboratories are, of course, different from artificial laboratories because they are influenced by a variety of uncontrollable variables and must face the challenges of adapting concrete implementation of policies to complex socioeconomic environments.

In chapters 1, 3, and 4 we highlighted the ways that municipal governments have, through policy and the dedication of resources, created institutional spaces to encourage diverse urban actors, such as residents, entrepreneurs, researchers, and civic organizations, to seed social innovations and local governance solutions to the delivery of housing, food, mobility, and other goods and services. These spaces for experimentation are place based and can seed social entrepreneurship and governance innovations. They can range from neighborhood laboratories and urban innovation hubs to new city agencies and public policy collaboratories. The challenge is often to scale these examples across a city.

The Amsterdam Smart City platform project is an example of how this experimentalism can scale. The project encompasses about 150 pilots across the city, supported by a few urban living labs spread across different neighborhoods (Neieu-West, Ijburg, Marineterrein, Buiksloterham, and Arena Stadium). These labs are designed to stimulate, coordinate, and support the pilots as they are developed and applied. A key factor for the project's success in piloting different applications of Amsterdam's approach to a smart city is the presence of an administrative organ and official position within the city administration, the Chief Technology Office/Officer (CTO), whose role is to merge scientific rigor with policy design. That office is also charged with ensuring that the pilots are conducted through inclusive, collaborative, and place-based processes, even if they slow down the implementation of some pilots that might otherwise have gotten off the ground more efficiently in a more centralized process (Karvonen and van Heur 2014, 10).

A similar organizational innovation that supports urban experimentalism is the Chief Science Offices, which many cities such as Amsterdam and Reggio Emilia have created and that employ PhD students, researchers, and other professionals to help craft evidence-based policies and adopt robust empirical and applied research techniques to organize pilot projects. These offices could interact and cooperate to develop an interoperable methodological language to address the lack of a standardized body of methods and techniques on urban experimentations. The role of this kind of Chief Urban Science Officer might be seen as the point of connection between the two design principles that depend on the local government—*urban experimentalism* and *enabling state*.

The use of different types of spaces for urban experimentation is a positive development, as others have noted, signaling a new era of urban innovation (Karvonen and van Heur 2014). These spaces help local governments overcome significant barriers to innovation, such as the trap of excessive bureaucracy (da Cruz et al. 2018, 4) and risk aversion (Sørensen and Torfing 2011). When the focus is on co-governance, adaptiveness becomes a feature of the system in that the participation of various actors is most often provided on a voluntary basis, compared with more hierarchical and vertical arrangements typical of structured and unitary organizations in which actors have stronger incentives to maintain the established governance system and resources' investment (Emerson et al. 2012, 19). Of course these spaces of innovation carry their own sustainability challenges, given that they are dependent on the political will of the current administration and face capacity limitations if they are structured to on a coordinative role by civil servants that might lack the necessary skills to manage a complex process involving multiple actors that have no history of interacting (Nesti 2018). However, increasingly local governments are experimenting with these spaces of innovation and consulting with academic and other partners to increase their capacity to lead co-design processes that result in policy and project prototypes that can be tried and tested in different neighborhoods.

One of the results from the Bologna experiment to become a *collaborative city* was the city administration's creation of the Office for Civic Imagination to help coordinate the neighborhood laboratories operating throughout the city. This move by the city was the result of an administrative decentralization reform approved by the city that changed the function of the city's neighborhood councils by putting them in charge of stimulating cooperation between residents through the development of different projects in their communities. This new function of the councils explicitly included the mandate to "work to promote the sense of territorial community, the culture of proximity, solidarity and collaboration between individuals and between city organizations, also according to the setting of community network and of shared administration that is based on the principle of horizontal subsidiarity as per article 118, last paragraph, of the Constitution" (Article 5, Reform of Neighborhoods, Annex A to P.G. No. 142306/2015).

In the Pilastro neighborhood project, the city was able to implement complex policy experimentations interconnecting the three possible dimensions of the neighborhood as a space of civic engagement, sustainable economic development, and regeneration of urban commons. As explained in chapter 3, this led to the creation of the Pilastro Northeast Development Agency centered on the role of neighborhood institutions and cultural and social values of the district identity. In other neighborhoods, Bologna adopted a different approach, more adapted to the specific circumstances and needs of the context. In the Bolognina neighborhood, for instance, the city initiated in 2014 a process of participatory urban planning codified into a *pact of co-living*—Convivere Bolognina (Comune di Bologna n.d.). The pact identified neighborhood priorities: efficient waste collection and reuse for local commercial activities through recycling and circular economy; increased revenues for local commercial activities through renovation of the urban public space, such as sidewalks regeneration; improvement of the broadband access; valorization of cultural diversity (the Bolognina neighborhood is the neighborhood in Bologna with the highest number of foreign residents, 24.5 percent compared to the city average of 14.5 percent) (Comune di Bologna n.d.). The overall goal of the pact is to coordinate the actions of existing projects that NGOs, local shops and craftworkers, and groups of city residents are carrying out to address these priorities as well as to stimulate new projects that these actors can implement in collaboration.

Within these broad programs, a co-design urban laboratory was carried out as one of the fieldworks of the Co-Bologna project. The lab involved public actors (i.e., public real estate managers at the national and city level [Federcasa and ASPII Bologna]); NGOs and representatives of local commercial activities and businesses (Kilowatt, Hotel/Restaurants, and Guercino, a civic association of local commercial shops in the Bolognina neighborhood); and informal groups of residents. The co-design lab's goal was to work on the specific co-creation project and the provision of collaborative services in private shared spaces within condominiums and public housing compounds, and then create synergies and collaborations with NGOs and local commercial spaces to extend those services to public spaces and generate new services that can serve the needs of other neighborhood actors. Examples of the kind of collaborative services they hoped to provide were shared maintenance of courtyards or shared entrances or the creation of community gardens. The lab led to the identification of priorities, synergies, and structured collaboration opportunities between the neighborhood actors involved as well as to the creation of a Community Association of Neighborhood shops of Bolognina (Co-Bologna n.d.).

Similarly, in Reggio Emilia the *neighborhood-as-a-commons* policy mentioned in chapter 3 made adaptability a working method to build different institutions from one neighborhood to the other. One of the most successful projects that evolved from this process was the Coviolo Wireless initiative which has successfully developed a broadband infrastructure in an underserved neighborhood, extending broadband access that enables social and economic development opportunities. The project turned neighborhood community centers into hotspots and managers of this digital infrastructure. The Coviolo initiative, first seeded in two neighborhoods (*Massenzatico* and *Fogliano*), is now being expanded to four other neighborhoods (*Cadè, Cella, Masone,* and *Marmirolo*) and will be implemented by the community that will manage the digital infrastructure (Comune di Reggio Emilia n.d.; JRC Science Hub Communities 2020a, 2020b).

The third feature of urban experimentalism is an *iterative process*, which entails creating and adopting a methodological approach to call forth collaborators and partners in the co-design process to deliberate, practice, and arrive at adaptable practices and policies. In order to be sustainable, experimentalism in the city context must find ways to be transferable and scalable across and within local contexts. In our own work within cities, we have applied the *co-city protocol*—a process or cycle that has the capacity to adapt to different places and can be applied across contexts to establish best practices for collaboration and cooperation between many kinds of actors (Baccarne et al. 2014).

The co-city protocol includes six key phases: knowing, mapping, practicing, prototyping, testing, and modeling. The first phase of the protocol is the *cheap talking phase*, which first emerged in game theory (Farrell and Rabin 1996) and later was applied in research on common pool resources (Poteete et al. 2010). In this phase, participants identify informal settings that allow for face-to-face and pressure-free communication among key local actors to activate the community of stakeholders that will be involved in the collaborative project. These discussions or sessions are organized in a variety of settings with significant outreach done in the local community, often through anchor organizations.

The second is the *mapping phase* and involves understanding the characteristics of the urban or neighborhood context through surveys and exploratory interviews, fieldwork activities, and ethnographic work. In this phase it might be necessary to create a digital platform as a tool for disseminating information and engaging various communities and stakeholders by co-creating a visual representation of potential urban commons through the analysis of relevant civic initiatives and self-organization experiences. The aim is also to understand the characteristics of the specific urban context and being able to design and prototype the kinds of governance tools and processes to be used later in the cycle.

The third phase, the *practicing phase*, is designed to identify and create possible synergies and alignments between projects and relevant actors. At the heart of this phase are co-working sessions with identified actors who are willing to participate in putting ideas into practice. This phase might culminate in a collaboration day or collaboration camp that takes the form of place-making events—for example, organization of cultural events, temporary use of abandoned building or spaces, or micro-regeneration interventions using vacant or available land or structures such as the creation of a neighborhood community garden—to prepare the actions for start of the co-design process.

The fourth phase, the *prototyping phase*, focuses on governance innovation. In this phase, participants and policymakers (local officials) reflect on the mapping and practicing phases and begin to extract the specific characteristics and needs of the community that will be served. One goal of this phase is to verify the conditions that promote the establishment of trust within the community and with the external actors. It is in this phase that the specific policy, legal, or institutional mechanism is co-designed to solve the issues and problems identified in the previous phases.

The penultimate phase is the *testing phase*. In this phase, project and policy prototypes are tested and evaluated through implementation, monitoring, and assessment. Both qualitative and quantitative metrics are employed to assess whether implementation is consistent with the design principles, objectives, and outcomes identified in earlier stages. This phase is often performed working with one or more knowledge/academic partners to design appropriate indicators and metrics to capture the desired outcomes and impacts from the project.

Finally, the *modeling phase* focuses on adapting and tailoring the prototype and nesting it within the legal and institutional framework of the city or local government. This phase is realized through the study of relevant legal laws, regulations, and administrative acts and through dialogue with civil servants and policy makers. This is an experimental phase involving perhaps the suspension of previous regulatory rules, the altering of bureaucratic processes, and the drafting of new policies that might also have a sunset clause and then a re-evaluation period. It can also involve the establishment of external or internal offices or support infrastructure in the city to support the new policies.

PRINCIPLE 5: TECH JUSTICE

The final principle of tech justice highlights access, participation, co-management and/or co-ownership of technological and digital infrastructure and data as an enabling driver of cooperation, collaboration, and social cohesion (Iaione, De Nictolis, and Berti Suman 2019). Technology can provide crucial tools for communication, to connect actors, and to facilitate the pooling of resources and actors. Access to digital devices and platforms, and to broadband, is also critical to addressing urban inequality as it can generate economic opportunity and facilitate access to essential goods and services, such as job opportunities and educational resources. Without connectivity and the ability to communicate, it is impossible to realize one's goals, to flourish, and to connect to others and build social capital across economic and cultural lines. As we have seen in some of the examples in chapter 3, open digital infrastructure can generate a virtuous cycle of openness, innovation, more investment in urban digital infrastructure that brings needed benefits to vulnerable groups (Sylvain 2016).

The lack of access to technology, particularly for underserved populations in many cities, is increasingly being addressed through innovative digital commons like mesh networks and community-based broadband networks. Mesh networks have been established in many European and US cities, including the famed community mesh network in Red Hook, Brooklyn, designed to overcome the digital divide, the Detroit, Michigan, network that provides connectivity to the 40 percent of its residents without internet access, and the mesh network in Berlin, Germany, designed to provide vital internet service to newly arrived migrants living in refugee shelters. Among the well-known and celebrated examples of wireless or metropolitan area wireless networks in the EU are the Spanish Guifi, the Greek AWMN, the Italian project Neco, and Ninux and Freifunk in Germany. Many of those are considered a democratic re-appropriation of technology (De Filippi and Tréguer 2016). Community mesh and broadband networks also promote what legal scholar Olivier Sylvain calls *broadband localism*, an approach that seeks to overcome broadband infrastructure and service disparities by race, ethnicity, and income (Sylvain 2012).

The next level of these constructed digital commons is illustrated by a community-based *edge-cloud* broadband network currently being designed and tested through participatory protocols in Harlem, New York City (Foley et al. 2022). The Harlem community, like many other ethnic minority urban communities, is facing obstacles that extend beyond broadband access and include the entire home, office, and IoT/smart city technology ecosystem. Although New York City is a *smart city*, it faces a stark *digital divide* that leaves one-third of households and families without access to broadband internet at home. Beyond the edge cloud, the project includes development of low-cost keyboard, video, and mouse (KVM) devices that will be used by a diverse set of community members to establish performance metrics for the edge cloud and identify system usability by the community, especially as it relates to closing the digital divide. These disaggregated devices open the potential for low-cost, secure user devices that are governed by a shared, centralized IT management team that oversees a high-performance edge cloud accessible to everyone in the community. Projects like these, which affirmatively further distributional equality in internet access, are part of a movement to go beyond *network neutrality* to *network equality* (Sylvain 2016).

Many of the design principles applied by these community networks resonate with the tech justice design principle. As adopted in the Declaration of Community Connectivity, these include (1) collective ownership (the network infrastructure is owned by the community where it is deployed); (2) social management (the network infrastructure is governed and operated by the community); (3) open design (the network implementation details are public and accessible to everyone); (4) open participation (anyone is allowed to extend the network, as long as they abide by the network principles and design); (5) free peering and transit (community networks offer free-peering agreements to every network offering reciprocity and allow their free-peering partners free transit to destination networks with which they also have free-peering agreements); and (6) the consideration of security and privacy concerns while designing and operating the network. The Declaration was facilitated by the UN Internet Governance Forum (IGF) Dynamic Coalition on Community Connectivity (DC3) (Belli 2016; Weinberg et al. 2015).

Another important facet of tech justice is the creation of digital platforms that enable residents to play an active role in shaping public policies, sometimes by voting or otherwise registering their preferences. These platforms can empower early and meaningful participation in the co-creation process by allowing participants to propose ideas and begin working collaboratively on the development of solutions. They can also enable the introduction of learning pathways and capacity building for residents who may not have the means to undertake specialized education but are able increase their knowledge and skills through well-thought-out collaborative design processes.

For example, in Barcelona the administration of Mayor Ada Colau has created a powerful digital infrastructure, *Decidim Barcelona*, for public consultations, resulting in increased transparency and participation in creating the policies and taking part of the activities of the city. Another example is the city of Paris, which utilizes online deliberation and voting as part of its participatory budgeting process. The city of Athens has also developed a platform, *SynAthina*, to facilitate urban co-governance partnerships. The platform acts as a networker and coordinator: residents, NGOs, and civic groups can submit ideas for voluntary activities in public spaces or other ways to utilize urban assets in collaboration with relevant government representatives, NGOs, and private actors. Online applications and discussions continue with offline meetings organized in a physical space, the SynAthina Kiosk. The platform and the meetings currently host thousands of users organized by groups of urban residents in cooperation with various private or civic sponsors.

Increasing the digitization and accessibility of democratic, collaborative processes at the local (regional and municipal) level is also exemplified by the process in the city of Bologna and its efforts to become a collaborative city. The Bologna *Iperbole* digital platform has functioned as a dissemination platform, allowing its collaborative and participatory processes to be widely known. Through the Collaborare è Bologna storytelling campaign, the city shed light on the implementation of the Bologna Regulation. The digital platform allows users to observe and participate in various experimental processes developed in urban co-governance field labs in three neighborhoods. The platform's section on *experiment with us* enables interested parties to pursue new projects or experiments in the city and to share their experiences doing so. As described in chapter 4, the city of Turin is implementing a similar, even more innovative platform, applying block chain through the First Life platform as part of its co-city experiment.

The principle of tech justice, which is a feature of so many of the neighborhood-based projects and citywide policies that we surveyed as part of our research, in the end analysis has many dimensions that are capable of application in specific contexts. These dimensions are (1) access and distribution; (2) broad participation; (3) co-management or co-governance of the platform; and (4) co-ownership of digital resources or data. It is rare that all these dimensions are present in a project or policy. Instead, we might view these dimensions as steppingstones toward the establishment of digital common resources consistent with the co-cities framework. These dimensions can also be used as metrics to steer the development of a smart city architecture towards a more just and democratic city. The Co-Roma.it is an attempt to embed these dimensions in the design, implementation, and management of a just and democratic smart city platform (Iaione 2019b).

Scholars of the urban commons have devoted little to no attention to the disruptive impacts of technological development on urban governance and city inhabitants' right to participate in the development of the city, and how technology can enhance the protection of human rights in cities. The smart city model, adopted by cities all over the world, presents a unique opportunity to innovatively tackle significant urban problems while reinventing the city in a more open and innovative form through more distributed data and technological capacity. However, the idea of the smart city as strongly aligned with sophisticated smart technologies faces the risk of increasing inequalities by stressing the gap between *haves* and *haves not* and deepening social divisions. What the tech justice principle does is to recognize the technological innovation embedded in the smart city model but then to shift attention away from the needs of the market for those technologies and toward utilization of them to leverage human and social capital to *open up* the potential for the application of smart technologies to address a range of socioeconomic and ecological challenges in cities (Deakin 2014, 7). Each dimension of our tech justice principle can push smart city protocols toward a city that reflects the right to the city, recognized as the right of every human be a part of the creation of the city and the stewardship of its shared resources.

The desire to leverage the assets of the smart city model to empower ordinary citizens, particularly those on the margins of our cities, is also why *co-management* and *co-ownership* are the highest dimensions that characterize the tech justice principle. These dimensions signify whether, as result of full access to technology and the overcoming of the urban digital divide, communities involved can collectively participate in and construct (or co-create) their own cooperative digital platforms and resources. This dimension is also concerned with the ability of residents to utilize those digital platforms and resources to acquire and develop skills that enable microenterprises or civic digital enterprises that support local economies.

This is what New York City was up to when it launched NYCx Co-Labs, described in chapter 4, dedicated to improving service delivery, spurring economic growth through new *civic tech jobs*, and increasing digital inclusion for all New Yorkers. NYCx Co-Labs was designed to offer a unique opportunity for residents to develop civic technology skills, collaborate on the strategic identification of community needs, and apply their newfound tech knowledge to co-creating solutions to local problems. By co-developing impactful technologies alongside civic technologists, startups, tech industry leaders, and city agencies, residents can increase their knowledge of civic technology and capacity for leadership and entrepreneurship. The initiative, as initially conceived, was also committed to community co-ownership of the systems that they help to research and develop. Engaging in co-development partnerships with the technology industry and community stakeholders' surfaces questions about how to best manage the intellectual property and technology transfer implications in a multistakeholder enterprise. As the co-labs program progresses, the team anticipates exploring mechanisms to ensure fairness in calculating value that is created collaboratively and is shared equitably. They are studying mechanisms like joint-venture agreements, accelerator-inspired equity models, distributed ledger systems, and community land trust agreements, and their potential applicability to our programs.

Another notable example is the city of Barcelona's shift toward *technological sovereignty*, which aims to rewrite the smart city agenda for city residents to embrace the right of the public to their information and data as well as to grant the public a right to open, transparent, and participatory decision making through new digital and platform technologies (Ribera-Fumaz 2019).

The Sidewalk Labs' proposal to establish an independent civic data trust that would control and govern all urban data as part of its Quayside Waterfront smart city project in Toronto has been proposed as a further attempt at digital sovereignty. While Sidewalk's interest in a data trust has provoked an intense curiosity about the idea, the privacy concerns raised about the tool and the failure of the project means that it will have to be tested elsewhere (Goodman and Powles 2019). Nevertheless, the project raised the possibility that guided by urban authorities, urban citizens could produce, access, and control their data and exchange contextualized information in real time through institutional co-governance platforms that could ensure confidentiality and accountability. On a practical level, a data trust has the potential to empower urban communities by giving them control over the knowledge on their potential and existing users, which allows them to provide services that are responsive to their needs.

Data trusts can exemplify tech justice because they give communities negotiating power against privately owned platforms. An open, democratically controlled, and collectively owned data trust is attractive for users who might perceive innovative data ownership models as carrying a higher level of protection against privacy concerns (Mills 2020) and prevent the exploitation of their data for marketing purposes. Communities can also benefit by managing data as the object of governance (i.e., in the case of a platform) or as a tool to provide a service or manage an urban common because they would have full control of economic revenues and the value produced using their data. Because the underlying technological infrastructure on which tech companies rely is often publicly funded and the data that makes these businesses profitable is collectively produced, economist Mariana Mazzucato has argued for the creation of a public repository that could sell data to companies rather than the other way around (Mazzucato 2018a).

CONCLUSION
NEW CO-CITY HORIZONS AND CHALLENGES

The design principles articulated in chapter 5 are based on projects undertaken, surveyed, and studied. However, much of what is exciting about our research and applied experiments in different cities is the potential for the co-city approach to spread and scale beyond the mostly European and US cities featured in our survey results to other continents and thus to other political, social, and economic contexts. Although we identified projects and plans all over the world that have some of the indicia and beginnings of a robust co-city approach, many if not most of these lacked support from local governments or financial investment sufficient to become transformative for urban residents. On the horizon are new challenges from projects in Baton Rouge, Louisiana, and Rome, Italy, which will test the power and saliency of the co-city approach to address endemic racism and injustice in an American city and bureaucratic ossification and wealth concentration in a capital city with one of the richest cultural heritages and most vibrant sustainable innovation ecosystems in the world. These applications raise new opportunities and challenges along with new research questions that we will briefly reflect on as we close out the book.

Looking to the future, we detect seeds of a commons-based approach in parts of Africa, Asia, and Latin America that are characterized by efforts to create new forms of co-governance, co-ownership, and stewardship of infrastructure, goods, and services that serve the most disadvantaged communities. New commons-based institutions and economies in African cities include the spread of co-created energy communities in African countries like Cameroon, Kenya, South Africa, Uganda (Ambole et al. 2021), and wireless community networks in cities like Cape Town and Pretoria, South Africa. These include the creation of a network of Fab Labs collectively owned by their members which develop alternatives to waste management, mobility, and economic resources. The Labs prototype, build, and test digital tools that can provide business opportunities for the community residents. In 2013, Woelab (Lomè, Togo), which makes technology accessible to all in the community, collectively constructed the first 3D printer built in Africa (Osayimwese and Rifkind 2014). The Ker Thiossane project (Dakar, Senegal) has created a park, a Fablab, and a School of the Commons and organizes many artistic and cultural interventions in the neighborhood.

In Asia, the emergence of Urban Villages in cities like Seoul and various Chinese cities resonates with many co-city principles and with collective-economy-based organizations. Urban villages can emerge from the effort of local officials or from community and/or private initiatives.

For instance, the Seoul Metropolitan Government supported the creation of the Seoul Community Support Center (SCSC) and the *Village Community Movement* (VCM) as forms of community-based economic development at the neighborhood level. The city invests in facilitation of community building to stimulate the creation of cooperatives that deliver services that could make the village self-sustainable, such as food coops, preschools, or co-housing buildings. The inspiration for the creation of urban villages likely derives from the experience of the Sungmisan village, the first urban village to be created in Seoul. This village was founded in 1994 when a group of neighbors joined forces to set up a local childcare center. Today, it includes about seven hundred households as well as seventy businesses and other institutions and runs a Village School and a consumer coop for purchasing ecofriendly goods (Bernardi 2017b).

Other examples of self-sustainable communities built as urban villages can be found in China within large metropolis areas such Shenzhen and Guangzhou (Tang 2015; Chung and Unger 2013). China's urbanization is characterized by the territorial expansion of cities primarily through the expropriation of surrounding rural land and its integration into urban areas. Urban villages in China emerge when rural villages are geographically incorporated in cities and granted urban administrative status. Chinese urban communities are typically governed by local residents' committees, which are part of the state governance system and responsible for delivering public services. In urban villages, however, it is the village shareholding companies that play a leading role in local governance. The shareholding companies are supposed to focus on economic activities only; that is, renting out collectively owned land or buildings to local factories and investing in real estate or services businesses. However, they also actively participate in community governance and look after the villagers. They provide welfare programs and other community services, sponsor community activities organized by residents' committees, and mediate conflicts between residents. The future of this governance depends on whether and to what extent the urban villages manage to maintain their collective assets and develop their village collective economy. The stronger and more profitable the village collective economy is, the more governance autonomy the village is likely to sustain (Tang 2015).

In Latin America, community land trusts in informal communities and favelas are being adapted to maintain self-constructed communities and to avoid gentrification in areas that are on the periphery of cities like Rio de Janeiro. According to organizers in Brazil, the basic logic of CLT governance and the idea of land stewardship are fitting for favelas that are characterized by residents who can own and sell their homes through an affordable housing market but yet do not own the land on which those houses sit (Williamson 2018, 17–18). That land is, in a sense, owned collectively, and residents' associations and other neighborhood institutions collectively govern the community by engaging in and advocating for infrastructure improvements in the community. However, collective governance of the favelas is precarious because of the tenuous authority that local authorities have over the favelas. Establishing a CLT would formalize this collective governance, represent the community, take action to improve that land, and provide security from eviction and real estate speculation.

Unlike CLTs in the US and Europe, which operate as nonprofit developers of land, CLTs in informal settlements and favelas would serve more to formalize existing housing and community stock. It would require that current homeowners *opt in* to the CLT, allowing the CLT to hold their titles, in exchange for lower property taxes and long-term affordability (Williamson 2018, 18–21). Communities in places like Rio are drawing inspiration from the CLT in San Juan, Puerto Rico, described in chapters 1 and 2, which includes seven communities that had built five thousand homes informally along the Martín Peña Canal in one of the most densely populated areas of Puerto Rico.

These examples from African, Asian, and Latin American cities intrigue us. We will continue to investigate them, and others, for how strongly they reflect the presence of the co-city design principles. At the same time, we continue to develop and refine these principles and the co-city approach developed in chapter 5 through application in different urban contexts, working with local partners and a constellation of knowledge institutions, public officials, and private enterprises. We describe two of those projects below and the ways they are challenging the co-city approach, on the one hand, while expanding its reach.

ADDRESSING STRUCTURAL RACISM: CO-CITY BATON ROUGE

The Black Lives Matter (BLM) protests around the world have drawn attention to the structural nature of racism and its devastating effects on Black communities. Nowhere is the legacy of racism and discrimination more evident than in the US, where this contemporary movement was born. More than fifty years after its historic Civil Rights Movement, and federal desegregation efforts, the US continues to suffer from persistent and deep racial segregation. This racial segregation and geographical stratification are particularly notable in US cities of all sizes with significant African American populations. This segregation is partly a legacy of legally segregated neighborhoods from the mid-twentieth century Jim Crow era and also from continuing discrimination in housing and financial lending, as well as the enduring racial preferences of whites choosing to live near other whites (Rothstein 2017). Many segregated American cities are the product of mid-century *White flight* to newly built suburbs, which excluded Blacks, in order to resist the racial integration taking root in urban schools and neighborhoods. The resulting infrastructure decay and capital disinvestment in these areas over the past forty-plus years have underscored the need for revitalization and equitable development in these urban neighborhoods.

Unfortunately, the US federal government's *urban renewal* and urban revitalization policies, as well as local efforts, have historically left these communities short of economic rehabilitation and often did more harm than good. Although urban renewal policies and practices have shifted over the decades, leading to important distinctions between the older mid-century efforts and latter twentieth-century and more contemporary efforts, one recurring pattern between the two periods is the focus on stimulating the redevelopment of underutilized areas located near central business districts across the country (Hyra 2012). The result has too often been displacement of Blacks, particularly the poorest, from central city neighborhoods with rising land values, and simultaneously the abandonment and persistent disinvestment of Black neighborhoods on the periphery of the urban core. Because of this history, many Black urban communities are deeply distrustful of any top-down policies and planning solutions that have not empowered their residents or community-based institutions.

The widespread economic inequality and deeply rooted, persistent racial segregation in America provides an opportunity to demonstrate the transformative impact of the co-city approach. For this reason, the application of the co-city model in the US has focused on cities and communities that continue to feel the effects of the historical legacy of racism for far too long. The first application of the co-city protocol in the US was used in Harlem, New York, to address the *digital divide*—that is, the inequitable access to the internet of low to moderate-income, residents of color—in an otherwise "smart city." The project sought to leverage the example of user-created and collectively managed wireless "mesh" networks created in different cities in the Europe and the US, and principles of digital stewardship, to sketch a co-created, community-managed network computing environment (Foley et al. 2022). The lessons from that project—specifically the pre-conditions for collaboration, which include trust-building and power asymmetries between stakeholders—have been carried over to a project with a set of broader challenges in a different urban context but with similar legacies of historic injustices and contemporary inequities in access to essential goods and services.

In 2019, the Co-City Baton Rouge (CCBR) project launched by partnering with a local redevelopment authority, Build Baton Rouge (BBR), and their mission to revitalize the historically African American Plank Road

Corridor of Baton Rouge, Louisiana. Plank Road was once a thriving commercial corridor but over the years has suffered from the white flight and disinvestment of similar Black communities in the US as well as from deeply flawed urban renewal practices. The City-Parish of East Baton Rouge (the city), where Plank Road is located, has a population of approximately four hundred fifty thousand and demographically is about 50 percent white and 50 percent African American. The surrounding urban metro region, referred to simply as Baton Rouge, is the capital of Louisiana and has a population of approximately eight hundred thousand.

Baton Rouge is spatially segregated by race and income, in what some describe as a tale of *two cities*, with higher-quality housing, amenities, and transportation in white areas and a lack of these amenities in Black areas, one of which is Plank Road. Plank Road extends for over four miles and varies in the character of its built environment. Sidewalks, although present, are inconsistent and not continuous. The Corridor is bordered by mostly commercial land uses, with residential uses in the intersecting side streets and extending for several blocks in either direction. The northern end of the Corridor contains more established businesses and is considerably more developed, whereas the southern and middle portions of the Corridor are considered severely *blighted*, with hundreds of vacant lots and dilapidated buildings. While many of the city's social and economic challenges are concentrated along this Corridor, Plank Road is also a significant anchor for the surrounding neighborhoods because it contains numerous assets. These assets include not only the available land and buildings that hold the opportunity for productive reuse but also strong social and civic organizations and institutions.

Over the last thirty years, there have been many failed efforts to revitalize the area. However, a 2019 Plank Road Master Plan has opened the door for a different approach to revitalization through extensive community engagement. Inspired by this master planning process, which envisions the installation of a ten-mile Bus Rapid Transit line that runs through the city that will economically transform the project area, BBR began to institute the highly iterative and deliberative co-city protocol and adapt it to this community and city. The collaborative approach of the protocol was designed to build on the work done during the master planning process to develop and implement projects that are addressing the needs that the local community have articulated—specifically, affordable housing, accessible green space, access to economic opportunity, and local capacity building.

As a first step in breaking from the legacy of urban renewal efforts in communities such as Plank Road, the redevelopment authority rebranded itself in 2019 under its new leadership. Previously known as the East Baton Rouge Redevelopment Authority, the then-president and chief executive officer Chris Tyson wanted to position the organization to be more akin to an urban laboratory that engaged the communities with which it works to solve its challenges and shape development outcomes. Tyson initiated a four-month visioning process in which the organization (through a consultant) conducted surveys and met with residents and stakeholders across the parish to arrive at a new identity and strategic vision. Part of the new vision of the agency is to "advance partnerships to build community-wide capacity" and to "bring people and resources together to promote equitable investment, innovative development, and thriving communities" (buildbatonrouge.org). Tyson understood that the new agency name and mission could not be only a matter of branding but must represent a new approach to development and revitalization of local communities that have been historically deprived of investment and virtually ignored as potential agents of their own revitalization.

Working closely with BBR, the focus of CCBR is on increasing the capacity of this community to pool resources with other local actors that will enable residents to determine how best to govern the process of neighborhood regeneration. The approach to resident-driven revitalization includes promoting "community wealth building," a concept rooted in systems/network theory that creates an inclusive, sustainable economy built on locally rooted and broadly held ownership of community assets. CCBR is developing a portfolio of innovative co-designed and co-governed prototypes designed to prevent widespread gentrification, to create new kinds of community goods (such as housing, parks, and micro-entrepreneurial space), and to establish new institutions for community stewardship of these goods.

The CCBR prototypes to date include a co-designed Community EcoPark and a hybrid community land bank and trust (CLBT). These prototypes are made possible by the existing BBR Land Bank, enabled by state statutory authority, which has acquired and is assembling vacant land to be used for developing prototypes that meet a range of critical needs related to social and economic determinates of health and overall well-being. The newly created Community Land Bank and Trust, in particular, will support the community-driven development of land under BBR's control for housing, green space, and commercial uses.

The CCBR projects' focus on community-driven, co-created development is happening at a time when local governments are beginning to partner with African American communities as a way to address the legacy of systemic racism in the US. The City of Seattle, for instance, recently announced that it would transfer one million dollars and a decommissioned fire station to a local community land trust in a historically Black neighborhood, the Central District (Scruggs 2018; City of Seattle 2020). The grant from the city is designed to help the Africatown Land Trust develop affordable rental housing, homeownership, and business opportunities in the district. The fire station will be used to establish a Center for Cultural Innovation in the neighborhood, a collaborative effort between the community and the city's Department of Neighborhoods and Office of Planning and Community Development. These public-community partnerships are one way to address calls for racial justice, as the city of Seattle recognizes. "We at the City of Seattle understand the urgency behind making bold investments in the Black community and increasing community ownership of land in the Central District" (City of Seattle 2020).

The CCBR project is in its third year and has attracted significant financing from both local and national foundations. This new financing will support ongoing efforts to increase the capacity of community members to continue to engage in the design of these institutions and their governance structures as new investment flows into their community. Part of the co-city protocol, referenced in chapter 5, is mapping the assets—including material, social, institutional, cultural, digital, and others—that are available in a community before putting in place a particular institutional or policy prototype. As decades of community economic development literature has shown, identifying and leveraging existing community assets enables collective solutions by building on existing associations, organizations, relationships, and other local resources (Kretzmann and McKnight

1993). When paired with financial and technical assistance, even the most disadvantaged and marginalized communities can robustly participate in the creation and sustainability of neighborhood-based development and revitalization.

Ultimately, CCBR is thinking bigger, beyond the Plank Road area, to scale up and adapt the process and outputs across the Baton Rouge metro region. To do so requires that at the end of the project there will be an evaluation of the programs and policies that have been implemented. The evaluation will include both qualitative measures, undertaken through surveys of all participants of the process, and quantitative measures. The evaluations of the process will allow predictions of what policies and programs will be successful and what adaptations may be useful or necessary to increase the likelihood of successful interventions. Armed with the results and lessons of this evaluation, the hope is that the co-city approach can take root in other American cities that are characterized by persistent racial segregation and economic stratification, dis-invested and forgotten neighborhoods, with significant assets and resources (material and social) that can be leveraged to create new collaborative ecosystems that enable inclusive enterprises so that these communities can thrive.

SCALING UP: CO-ROMA

One open question often posed about the co-city model is whether, and how much, it can scale to global or capital cities. It is one thing to activate collective action around shared resources in specific neighborhoods, or focused on a specific resource (e.g., a park, a community garden, wireless network), or to launch a citywide collaboration policy of public-community partnerships in a place like Bologna, Turin, Naples, or Reggio Emilia. Although these are cities with sizable populations, they lack the administrative and political complexity of Rome or New York City, for example.

Rome as the capital city of Italy is a city of contradictions. Standing as the country's second most economically productive city, after Milan, it represents 9.4 percent of the share of the national GDP (UN-Habitat 2016). Yet, the City of Rome presents surprisingly elevated indicators of social and economic vulnerability, high unemployment rates, and significant disparities in wealth distribution and health conditions with large numbers of families in danger of falling into economic distress and diseases (ISTAT 2017; Lelo et al. 2021). Although Rome was at the heart of the Italian *commons movement*, it is also a city that has to date not been able to recognize or implement effective legal or policy protections for collective management or governance of its rich and varied community spaces.

The Co-Roma project was established in 2015, with the support of ENEA, the Italian National Agency for New Technologies, Energy and Sustainable Economic Development, and the Horizon 2020 OpenHeritage EU project. Co-Roma is designed to test whether and how the co-city approach can apply to a large metropolis. The Co-Roma project grew out of the university-based urban collaboratory, a concept that we described in chapters 4 and 5, to test the saliency of the design principles in relation to different types of urban commons, especially abandoned assets and underdeveloped infrastructure. The project has taken root through a focus on activating collective and collaborative governance of urban essential infrastructure and resources. Such infrastructure and resources include the city's natural resources (i.e., Tiber and Aniene rivers and several urban parks), energy provision, housing, culture, and heritage in distressed areas and neighborhoods of the city (Cellamare 2017).

Some milestones of the Co-Roma project thus far are the establishment of a coalition of social actors (Agenda Tevere), an agreement for the sustainability of the urban segment of the Tiber River, a community association established to safeguard the local cultural heritage according to the Faro Convention, a treaty recognizing the importance of local heritage to communities and society, and a multi-purpose neighborhood community cooperative (CooperACTiva). The cooperative will catalyze the development of three distressed neighborhoods' assets into sustainable tourism, urban farming, and communal energy enterprises.

In pursuing co-governance of the city's rich and varied community spaces, cultural assets have emerged as a key entry point, particularly in the most economically distressed areas. The Co-Roma project is seeking to implement the principles of the 2005 Faro Convention, specifically on the value of cultural heritage, through the establishment of a Faro Community (pursuant to article 2 of the Convention) and the establishment of public governance mechanisms that enable the "joint action" of community, public, social, cognitive, and private stakeholders.

CONCLUSION

The first major step toward co-governance in the project was to establish a foundation for the care and regeneration of the historic Tiber River, which runs through the city of Rome. The river is important to the cultural heritage of Rome, to which city inhabitants are strongly attached, and is also part of the critical green infrastructure of the city. The *Tiber for all* (Tevere per Tutti) foundation is the result of a process initiated by a coalition of civic, social, private, and scientific partners grouped within Agenda Tevere, a nonprofit association. The foundation was identified by the coalition as the most appropriate institutional tool to pool different actors and to raise the necessary resources to regenerate the river banks, promote the capacity building of local administrators on the issues of environmental preservation and sustainable development, support data collection and analysis on various uses and activities that take place on the river banks, and partner with knowledge actors active in the city (i.e., universities and applied research centers) to carry out field experiments exploring sustainable uses and regeneration activities of the river.

Due in large part to the efforts of the Agenda Tevere coalition, Sapienza University, and Luiss University (specifically the LabGov), the Regional Council of the Lazio Region approved a law (Lazio Region law no. 1 of 27 February 2020, article 20) mandating that the Governor establish the foundation. The Lazio Governor's office has already kick-started the implementation phase by passing a motion that activates the legal process to incorporate the foundation. In the meantime, thanks to the Agenda Tevere coalition, the first co-governance mechanism was established through the signing of the Tiber River contract in February 2022.

The second and most complex field experiment of the Co-Roma project, GrInn Lab, emerges out of the Luiss University LabGov's Urban Transdisciplinary Clinic which has been laying the groundwork for the field research program since 2015. This groundwork included attracting grants for the energy community project from the ENEA and the Horizon2020 program, as previously mentioned. LabGov also proposed to establish and implement a *smart co-district* as a way to implement the co-city principles and cycle in distressed neighborhoods around the city (Meloni et al. 2019). The initial phases of the co-city approach, including data analysis and mapping, revealed that the southeastern district of the city—where the Alessandrino, Centocelle, and Torre Spaccata neighborhoods are located—has the highest indicators of social and economic distress. These neighborhoods present very low Human Development Index values, with the lowest value and lowest income levels in the Torre Spaccata area (Lelo et al. 2018; d'Albergo and De Leo 2018; ISTAT 2017).

The university-based urban collaboratory seeks to revitalize the above-mentioned neighborhoods through the co-city cycle, described in chapter 5. The cheap talking and mapping (both analogic and digital) phases identified the key entry points for collective action as neighborhood identity, culture, and heritage. Each neighborhood contains significant cultural heritage and green infrastructure, including Roman ruins and major parks such as the Public Archeological Park of Centocelle that hosts two Roman villas, Villas Ad Duas Lauros, and Villa della Piscina (Gioia 2004), as well as the first military airport which opened in 1909. The park is currently located in the middle of a highly urbanized area, and it is only partially accessible to the public. It has never had great appeal either to tourists or to the local community due to the poor conditions of the area, which also host illicit activities that pose security threats to city inhabitants (Celauro et al. 2019). Other cultural assets include the Osteria di Centocelle, the historic Tunnel of Centocelle, and the Tower of Centocelle or San Giovanni Tower.

City agencies with responsibility for this district have failed to leverage the richness of this culture and heritage to develop and improve the neighborhoods within it. On the other hand, neighborhood residents and activists have coalesced into a movement to claim rights on, and to protect, this heritage. Through the co-city process, Luiss University's LabGov has helped to institute co-design laboratories, organized microregeneration activities (including the creation of community gardens in each neighborhood and placemaking activities to preserve the neighborhood's heritage, and other activities), and created a legal association for the collective action of local residents to care for their parks (the Community Association for the Public Park of Centocelle). The development of neighborhood labs has allowed residents to identify and focus their efforts on the establishment and sustainability of neighborhood collaborative welfare services, distributed energy production, and heritage-based sustainable tourism. The collaboratory processes also confirmed that the *co-district* is the most suitable scale to experiment with urban co-governance, given the territorial coalescence between the three neighborhoods (i.e., social, economic, and infrastructure) (Calafati and Veneri 2013).

Another significant Co-Roma development is the establishment of CooperACTiva, a multipurpose neighborhood community cooperative incorporated in December 2018. The community cooperative was established in order to leverage the social and economic pooling of efforts and resources that were already deeply rooted within the three neighborhoods. The business plan for the cooperative is based on three kinds of investments. The first requires a minimum initial investment and generates revenues from the sale of sustainable tourism services. The services developed by the community enterprise include citywide bike tours, electric mobility services at the neighborhood level, and a local food-based heritage promotion activity. The second is real estate investment and urban farming. The cooperative is exploring the acquisition of a large piece of land in order to revitalize and manage an existing urban farm. The third investment is to create an energy community from existing community resources—human skills, infrastructure, and available equipment—that would serve as a local node of a large network of energy communities. The by-laws of the cooperative contain a reinvestment clause that ensures that the district benefits from the pooling of human capacity and physical infrastructure and assets, and not just a small group of the most active residents and activists. This by-law clause requires that 30 percent of revenues must be reinvested in projects for the improvement of the neighborhoods even if they benefit residents who are not members of the cooperative. This allows for a collaborative economy, in which financial benefits flow more widely to sustain the various social enterprises and community-based activities of the cooperative.

Finally, Luiss LabGov established the social start-up GrInn.City to support the Co-Roma project through self-sustaining agriculture activities and the provision of adequate housing around the city. The agriculture work was inspired by the Luiss University community garden, which was the first field experimentation that LabGov researchers and students engaged in beginning in 2014. As part of that effort, LabGov students and researchers designed a governance plan to collectively maintain the garden space, co-managing it with university offices. This collaborative governance model, which students had to learn by practicing it, was later exported from the University into the city of Rome.

Luiss students created five satellites of the university community garden, leveraging their own fieldwork. The first one was developed inside a community space dedicated to children and their families in the Centocelle neighborhood, which is still used as an educational garden. The second one was created and donated to a Roma family informally living in a private space of the Centocelle Archeological Park. The third one was set up in Torre Spaccata, inside a public library, where people with disabilities (thanks to a local NGO) have the chance to manage the space. Each of these three community gardens was conceived with adaptive features based on the different urban context.

This field experimentation led students to deepen the study of urban community agriculture and food policy in Rome, which in turn has led to the establishment of two more gardens with educational missions—one in a primary school and the other in a museum (that studied the Luiss LabGov and Co-Roma management model to replicate it thanks to the funding of the Erasmus+ EU program). The creation of this network of schools, museums, neighborhoods, and university community gardens has led to a collaboration between Confagricoltura, the General Confederation of Italian Agriculture Industries, which represents and protects Italian agricultural identity, and LabGov for a research project on agriculture. Together, LabGov and Confagricoltura has been working on the creation of a digital platform devoted to the collection of data on food consumption and supply chains in order to make the agricultural industry and consumers aware of urban consumption trends and to thereby reduce food waste. The project involves the digital transformation of participating urban gardens in the city, using a system of small solar panels, to assist urban gardeners to collect and deliver data to the digital platform. Students and urban gardeners will be the first owners of the platform and the data. A connected strand of work will involve designing and refining the legal and governance aspects of community gardening in Rome, informed by lessons learned through the Ru:rban network project, to which LabGov students have contributed. That project funded was devoted to the creation and strengthening of a municipal regulation on urban gardens and agriculture (Karamarkos 2018). This regulation is considered to be the only real achievement for the commons movement in Rome, at least in terms of legal recognition of community rights to co-construct and co-govern shared infrastructure and assets.

A closely related strand of work of the start-up GrInn.City is devoted to self-sustaining adequate housing units. This strand too derives from the work of LabGov students, researchers, and practitioners and their experience gained over the years working on public and social housing policy. This work, done in collaboration with Federcasa, a federation of more than a hundred public agencies that build and manage public and social housing in Italy, has led to the development of a study that was partially synthesized in a book, *Housing for All* (Iaione et al. 2019). This work and experience will lead to the development of prototypes for different forms of self-sustaining housing units to be developed with Federcasa and the city of Rome. This collaborative housing project will be crafted using new technologies, such as blockchain technology, and new legal vehicles and instruments, such as "renewable energy communities," recognized by article 22 of the 2018 EU directive on the promotion of the use of energy from renewable sources. The business model will be based on the sale of the energy produced and will redirect surplus revenues toward other less profitable, but not less important, community welfare activities such as community-based healthcare services.

Last, the activity developed by LabGov in Rome led Luiss to become one of the main partners of the city of Rome within a major state-funded program: "House of Emerging Technologies." The program is part of an investment plan of the Ministery of Economic Development to support large broadband emerging technologies, starting with cities conducting 5G-related experimentations (Turin, Rome, Catania, Cagliari, Genova, Milan, Prato, L'Aquila, Bari, and Matera) (Mise Gov n.d.). The Rome project, named the Rome Open Labs, foresees the drafting of a 330 million euros worth Urban Integrated Plan for the Metropolitan Authority of Rome. The plan would sustain the innovation and climate change adaptation in the metropolitan city of Rome. The plan will also become the pillar of the Rome dossier for the candidacy to host the EXPO2030 exhibition. The Rome Open Labs will be physically hosted in freshly renovated spaces within the area of the Tiburtina train station to create a technology district in the heart of the city. The first project within the Rome

Open Labs creates a collaboratory in Rome with the city in cooperation with Luiss and three technical departments at other universities within the city (i.e., Sapienza Innovation, Tor Vergata School of Engineering, and Roma Tre Information Engineering Department) alongside several technical and corporate partners (Comune di Roma n.d.).

LOOKING AHEAD: OPPORTUNITIES AND CHALLENGES

Our involvement in the ongoing Co-City Baton Rouge and Co-Roma projects promises to yield new insights and challenges for the co-city approach. However, we are also mindful of the existing challenges that the co-city approach poses for future research and experimentation grounds. In closing, we identify five open questions or takeaways that will shape future applications of the co-city approach.

First, the co-city approach and its implementation are intentionally conceived of and designed to further inclusive city-making and to further social justice and racial justice. The most structurally disadvantaged and distressed neighborhoods and communities stand to benefit the most from our co-city approach. However, it is naïve to believe that any approach, including ours, is a panacea for what are fundamentally structural and systemic challenges in so many places around the world. The Black Lives Matter movement, and specifically the 2020 protests, underscore the depth of the challenge of institutionalized racism and inequality around the world, for example. Similarly, the headwinds faced by the commons movement in Europe, but specifically in Rome, has failed to gain sufficient formal and legal recognition over the last decade or so in large part because of vested interests entrenched in an old-fashioned urban development model.

These deep, structural changes face intense resistance all over the world. The approach that we propose here can be part of that change, but it too faces resistance because it requires disrupting current dominant economic models and social belief systems. Overcoming this resistance requires nimbleness and adaptability in applying the model. The most important lesson for us is, first, to listen to communities. At the same time, however, we must create bridges to other urban stakeholders and enable collective action, where possible, through inclusive and transformative deliberative processes. As we have argued in the book, this often benefits from the creation of public-community partnerships or public-private-community partnerships.

Second, there remains the challenge to attract to these urban commons institutions and co-city experiments financial institutions and private investors seeking social and sustainable investments rather than mere financial returns and profits. Although many of the projects, past and ongoing, have attracted significant state and philanthropic financial support, private sector investments too have the potential to become an enabling force. One hopeful sign is that financial investors are looking for impact to measure what they call "additionality" (European Investment Bank 2018) and "profit and purpose" (Fink 2019; 2020).

At the same time, we are also conscious of the risks of opportunistic behavior in the private sector. This implies that financial institutions nudge markets to advance social impact and environmental objectives such as climate-neutrality, climate mitigation and adaptation, sustainable use of resources such as water and marine resources, prevention and reduction of pollution, and protection and restoration of biodiversity and ecosystems.

Potential opportunities for responsible and transformative investment in co-city initiatives could result from the effort to develop an environmental, social, and governance (ESG) value proposition for private investment in financial markets. This effort is supported by emerging regulation, at least in the US and the EU, which would help steer financial investment away from environmentally harmful industries, possible misconduct issues, and potential governance failings. The European Union recently introduced Regulation (EU) 2020/852, known as the Taxonomy Regulation, which establishes an approach aimed at facilitating sustainable investment, as well as a platform for monitoring its implementation. The platform recently published a report on social taxonomy, which includes among its objectives the creation of "inclusive and sustainable communities and societies," which implies respect and support of human rights and emphasizing issues such as improving/maintaining the accessibility and availability of basic economic infrastructure and services like clean electricity and water for certain vulnerable groups or groups in need (Platform on Sustainable Finance 2022).

Third, commons-based legal and policy models are powerfully suited to guarantee accessibility and availability of basic economic infrastructure and services. Indeed, commons-based legal and policy models can introduce reinvestment clauses on urban infrastructure and services thereby guaranteeing that part of the value produced is captured by locals and residents. This legal and policy solution is the only one that can enable the "external mutualism" ingredient that ordinary cooperatives do not guarantee by design. We are aware of the legal and policy innovation that these models demand and also of the potential failures due mainly to the self-serving and opportunistic behavior that still animates many urban actors, including community representatives.

Fourth, we are aware that our analysis and our field research are based mainly on cases and experience taking place in developed countries, mainly the US and the EU. It is therefore crucial in the coming years to broaden the scope of the analysis and experimental work in a diverse set of countries to understand if and how the co-city approach can be adapted to contexts in which the quality of democracy or the institutional capacity is significantly lower and knowledge institutions enjoy fewer resources and funding opportunities. Communities and the civic nonprofit sector might be the driving force in these contexts. Knowledge actors and international organizations can be of assistance by redirecting or redesigning their cooperation or their cooperation programming in developing countries through the lenses of the co-city approach. An early attempt to work in Accra, Ghana, with the Ghana Institute for Public Management and a local NGO was a step toward identifying the potential and limits of the model in a developing economy in the Global South (Galizzi and Abotsi 2011). These efforts, by us and others, must continue there and in other parts of the world to demonstrate that the co-city approach can be constructed by observing, working in, and learning from a diverse set of cities. Only then can the co-city approach be proposed as a universally valid vision for a just and democratic self-sustaining city.

Last but probably the most important point of our personal intellectual journeys is that scientific, knowledge, education, and cultural institutions like research centers, universities, schools, museums, and libraries can play an important role by acting as brokers of strong alliances between individuals, the public sector, the private sector, and social or civic actors. They can help safeguard and protect community interests, acting almost like independent authorities to defend and represent the interests of the weakest voices. We have seen how they can serve the role of entrepreneurial, enabling platforms. They must also decide to boldly invest in the capacity building of individuals and communities in order for the kind of co-governance that we describe here to work in the most disadvantaged and marginal communities. We also need to accept the fact that not everybody in the community and not every community or neighborhood is ready or willing to be entrepreneurial, or be able and willing to create economically diverse systems through an urban-commons-based approach.

Equally important is that knowledge institutions can be the space where future generations can self-empower and equip themselves to join the fight for a just ecological, technological, and digital transition. Indeed, sustainable development implies an intergenerational solidarity: present generations shall act having in mind that they bear a duty towards future generations who hold a right to the future, which they share also with other species that have no voice or agency in the present.

The rights of future generations have already been codified in laws and even case law. For instance, according to the 2021 German Constitutional Court decision, Article 20a of the German Basic Law implies "the necessity to treat the natural foundations of life with such care and to leave them in such condition that future generations who wish to carry on preserving these foundations are not forced to engage in radical abstinence" (Federal Constitutional Court 2021). Similarly, in 2022, the Italian Constitution has been modified to follow in the footsteps of eighty-one other constitutions by establishing that the Republic (not just the state and therefore every individual, much like every social or territorial autonomy) must safeguard the environment, biodiversity, and the ecosystem "also in the interest of future generations." And this addition has been inserted in Article 9, the same article that establishes that "the Republic promotes the development of culture and scientific and technical research," and that it safeguards "landscape and the historic and artistic heritage of the Nation" (Italiano Legge Constituzionale 2022).

We believe that what is still missing is the design and implementation of policies that on a large scale can promote an intergenerational alliance to spread knowledge and take joint action about the daunting common challenges that the planet and all its species will have to tackle in the near future. The diffusion of this knowledge and collective action can help change the culture and behaviors of present and future generations and enlist their vast majority in implementing effective solutions to dangers that threaten the future existence of the Earth and all its species. At the same time, we are still missing procedures that give voice and agency to those that have no voice and no agency in the present, especially in its decision-making institutions and processes. Younger generations, other species, plants, and animals, and the not-yet-existing generations and species do not get to participate in the present electoral cycles that representative democracy periodically grants. Participatory, deliberative, associative, and collaborative democracy processes, tools and institutions need therefore to become increasingly more relevant if we want to safeguard future generations, as much as the future of the planet and democracy itself.

Thanks to the lessons learned through the co-city project, it is our strong conviction that intergenerational collective action, deliberation, and an alliance of minds and energies can take place or be accelerated within or with the support of knowledge institutions. It is also our firm belief that the presence of knowledge institutions can strengthen further public and social institutions' role in making sure that digital, technological, and ecological transition processes do not happen at the expense of the interests of the poor, the disenfranchised, those discriminated against, and therefore may contribute to building a more equal society and markets, as well as to the improvement of the quality of democracy by changing the way public institutions work.

The efforts developed and energies spent to research, experiment, and practice the co-city approach also made us realize that we need to build on the pillars of open science, citizen science, city science, and more generally responsible research and innovation for a new framework of analysis and engaged research that can guide younger generations willing to serve the interests of the planet, all its species, and their future generations in particular those that have no or reduced voice and agency. After all, the city is a wonderful classroom where everyone can learn and an unbeatable laboratory where everyone can experiment with new ideas and approaches. Therefore, we hope that the co-city approach will inspire present and future generations of researchers to initiate a co-science project that can transform cities into laboratories for experimentation and learning toward a just and democratic ecological and technological transition and therefore sustainable development.

APPENDIX

The database of Co-Cities represents the culmination of a six-year-long research project seeking to investigate and experiment with new forms of collaborative city making that are pushing urban areas toward new frontiers of co-governance, inclusive economic growth, and social innovation. The case studies gathered here come from different kinds of cities located all around the world; they include groundbreaking experiments in Bologna (Italy) as well as in other Italian cities (e.g., Bologna, Milan, Naples, Reggio Emilia, Rome), and in global cities such as Seoul (South Korea), Mexico City (Mexico), New York (New York), Barcelona (Spain), and Amsterdam (Netherlands).

This appendix presents an overview of the data set from over 140 cities that we investigated and analyzed (out of the over 200 cities surveyed). The data set provides several community-based projects and public policies from the cities mapped. All the projects and public policies presented in this appendix are also published on the web platform, Commoning.city. The intention behind Commoning.city is to provide an international mapping platform for researchers, practitioners, public officials, city agencies, and policymakers interested in understanding the variety of practices and policies that embrace the kinds of urban commons that we reference in the book. The first phase of the research, whose results are summarized here, was mostly exploratory. Some case studies were explored more in depth, including fieldwork observations and/or direct involvement in the case studies, as indicated by the data in the Exemplary Case Studies table at the end of this appendix. Our goal in creating the online dataset is chiefly to attract the interest of fellow researchers who could build on this first body of knowledge we offer here to further develop, improve, challenge, and rebuild the foundation of this line of research we humbly attempted to open up. The online dataset is open access, collaborative, and iterative. This means that we are constantly conducting further research to update the information on the projects and public policies we have surveyed so far, following their evolution. We continue to conduct research on a rolling basis to expand the dataset with new projects and public policies. Please, check the online version for updates and more information: http://commoning.city/commons-map/.

METHODOLOGY FOR DATA SELECTION AND DATA COLLECTION

The case studies have been extracted from different sources, including those listed here. The Co-Cities database, available on Commoning.city, indicates detailed source information for each case study. The sources include:

1. The papers presented at The City as a Commons conference in Bologna, Italy, in 2015. These papers contained many relevant cases and examples of urban commons in different geographic contexts. These papers are available in the Digital Library of the Commons or published elsewhere and thus are fully accessible;
2. Scientific journals covering the following themes: commons (e.g., *The International Journal of the Commons*) and urban studies (*CITY—Analysis of Urban Trends, Culture, Theory, Policy, Action*; *Policy Studies*; *Urban Policy and Research*; *Urban, Planning and Transport Research*; *Journal of Urbanism: International Research on Placemaking and Urban Sustainability*; *Journal of Urban Affairs*);
3. Academic conferences on the commons and urban commons and, in particular, involving urban research, cities, and policy studies. In addition to the City as a Commons conference in Bologna, examples include the 4th Conference on Good Economy; relevant thematic events on the commons and city-making (e.g., the New Democracy

workshops held by Pakhuis de Zwijger–Amsterdam; Sharitaly events in Italy; GSEF 2016—Forum Mondial de l'économie sociale; Urbanpromo conferences in Italy; Innovative City Development meeting in Madrid; the World Forum on urban violence and education for coexistence and peace held in Madrid; UNIVERSSE 2017—the 4th European Congress for Social Solidarity Economy held in Athens; and Verge New York City 2017 held at the New School);

4. Urban media (Shareable, Citiscope, CityLab, Cities in Transition, Guardian Cities, P2P Foundation, Remixthecommons, and OnTheCommons);
5. Direct suggestions from key experts, scholars, and practitioners: David Bollier, Silke Helfrich, Anna Davies, Marie Dellenbaugh, Fabiana Bettini, Thamy Pogrebinschi, Ezio Manzini, Eduardo Staszowski, and Martin Kornberger;
6. Deliverables produced by the EU research and funding program Horizon 2020—funded research project Open Heritage and EUARENAS;
7. In order to reach geographical areas not covered through the previously mentioned samples, we also engaged in internet data mining through established internet providers (Google and Bing) and scientific databases (Summon Discovery) using the following keywords: commons, urban commons, community land trust, Wi-Fi community network, collaborative neighborhood, collaborative district, collaborative governance, and community-managed services.

The cities that we investigated and surveyed were selected in order to provide us with a breadth of examples of different projects and policies of collectively or collaboratively managed or governend urban resources in different countries and contexts.

We identified and included a group of cities for every geographical area in order to capture diversity (although without any ambition of representativeness or statistical significance) of cultural, social, economic, legal, and institutional factors. The data collected from all cities is displayed on a map available here: http://commoning.city/commons-map/. For each project/public policy, a short record card has been uploaded on the commons map, including the main information collected through answers to the questionnaires and through online data mining, and through further research from scientific papers and industry specific magazines.

The record card uploaded on the website is built as follows:

City	[...]
Name of the Project/Public Policy	[...]
Date Initiated	[...]
Description of the Project/Public Policy	[...]
Urban Co-governance	[...]
Enabling State	[...]
Pooling Economies	[...]
Experimentalism	[...]
Tech Justice	[...]
Project Website	[...]
References, sources, contact person(s)	[...]

THE CO-CITIES DATA SET

The first mapping phase of the project resulted in a collection of 522 policies/projects in 201 cities in different geographical areas:

Region	Total cities	Total projects/public policies
Europe	90	306
North America	23	81
Central and Latin America	20	41
Africa	24	35
Asia	37	48
Oceania	7	11
Total	201	522

From this initial database, we more closely analyzed, through interviews with relevant stakeholders and/or more extensive desk research 140 cities with 283 projects/public policies (out of the initial 522 identified) within them. The cities that we surveyed and analyzed most closely were

APPENDIX

selected on the basis of the existence of a project or policy relevant to creating, enabling, facilitating, or sustaining collaboratively or cooperatively shared resources utilizing the existing infrastructure of cities.

Region	Total cities	Total projects/public policies
Europe	72	147
North America	15	41
Central and Latin America	10	23
Africa	13	29
Asia	25	36
Oceania	5	7
Total	140	283

CODING CITIES

The process for collecting the data contained in this report involved gathering information from secondary sources and/or contacting and interviewing a representative for each city mapped. This report presents a summary of the results of the empirical analysis carried out on projects/public policies in 140 cities.

For this stage of analysis we did not engage in a comparison of the collected case studies, which was planned for the second phase of the research after a larger number of projects/public policies were collected (in order to have good representation of all the geographical areas). In this report, the analysis of the 140 cities is strictly descriptive. Our aim is to highlight the relevant aspects of each city and to build a classification criterion for the four dimensions captured by the data. The charts and tables below present the aggregated results of the coding at a regional level and per city. 0 = absent; 1 = weak; 2 = moderate; 3 = strong.

The coding was carried out with research assistance and was guided by an analytical tool, the Co-Cities Guidance Codebook. In the Guidance Codebook, we operationalized each design principle and highlighted its main features on the basis of the literature review outlined in chapter 4. Every design principle operationalization is accompanied by a set of guiding empirical questions. The questions assisted coders in assigning a value to the design principle from 0 to 3. The Codebook counsels to assign values on an incremental scale, meaning that the greater the intensity of the design principle feature in the case study, the higher the value assigned. For example, the Tech Justice design principles are operationalized in 4 layers: lack of access to data/technology and/or absence of any involvement of communities in the tech management/ownership (absence); improved access to data/technology (low); collaborative management of the data/technology (moderate); cooperative ownership of the data/technology (strong).

REGIONS AND CITIES CODED AND ANALYZED

EUROPE

The European cities show on average an above-moderate score in the majority of the design principles considered in this study. With regard to Urban Co-Governance (2.3), Experimentalism (2.3), Pooling (2.2), and Enabling State (2.4), the European cities invested serious efforts in promoting public policies as well as projects. The European local authorities have been, on average, very active in promoting the urban commons and new forms of urban co-governance. The uniqueness of the European cases are the networks and frameworks in place between each city that create added value for each project. On the other hand, Tech Justice (1) is still an underdeveloped aspect of these cases, similarly to other regions, which signal the need for an expansion of the dataset to make sure it includes a wider number of cases concerned with technological, digital, data issues.

The dataset includes ninety cities with 303 projects/public policies and closely analyzed seventy-two cities and 147 projects/public policies.

Data aggregated per city

City	Urban co-governance	Enabling state	Pooling economies	Experimentalism	Tech justice
Aarhus	2	3	3	3	1
Amsterdam	2	3	2	3	2
Athens	2	2	2	2	2
Barcelona	2	3	2	2	2

Data aggregated per city (continued)

City	Urban co-governance	Enabling state	Pooling economies	Experimentalism	Tech justice
Bari	3	2	3	2	0
Battipaglia	3	3	3	2	0
Belgrade	2	3	2	3	2
Berlin	2	3	2	3	1
Bilbao	2	2	1.5	1	3
Birmingham	2	3	3	3	0
Bologna	2	3	2	2	2
Bristol	3	3	3	2	0
Brussels	3	3	2	3	1
Budapest	2	3	3	2	0
Callan	2	2	2	3	2
Caserta	3	2	2	3	1
Colombes	3	1	3	3	0
Copenhagen	2	3	1	2	1
Coruna	2	2	2	2	0
Dublin	2	1	3	2	2
Edinburgh	3	3	3	3	2
Eindhoven	3	2	2	3	3
Fidenza	3	3	2	2	1
Gdansk	3	2.5	2.5	3	1
Ghent	2	2	2	3	2
Glasgow	3	3	2	3	2
Gothenburg	3	3	3	3	2
Grenoble	2	1	1	2	2
Hamburg	2	3	2	3	0
Helsinki	2	3	3	3	2
Iasi	2	2.5	1	2	1

(continued)

Data aggregated per city (continued)

City	Urban co-governance	Enabling state	Pooling economies	Experimentalism	Tech justice
Lille	2	1	3	3	2
Lisbon	3	3	3	3	0
Liverpool	2	2	3	2	1
London	2	1	3	3	2
Lucca	3	2	2	1	1
Lyon	3	3	3	3	0
Madrid	2	3	2	3	2
Malmo	3	3	3	2	2
Mantova	2	3	2	3	2
Maribor	2	3	2	3	0
Marseille	2	3	3	3	1
Massarosa	1	3	2	3	0
Mataró	3	3	2	2	0
Matera	2	2	3	3	2
Messina	2	3	1	2	0
Milan	2	3	2	2	2
Montepellier	2	2	3	3	0
Narni	1	3	2	3	0
Nantes	3	3	3	2	0
Naples	2	3	3	2	0
Oslo	3	2	1	−2	1
Ostrava	3	3	2	3	3
Palermo	1	2	2	1	0
Padua	3	3	2.5	2	1
Paris	2	3	2	2	1

Data aggregated per city (continued)

City	Urban co-governance	Enabling state	Pooling economies	Experimentalism	Tech justice
Peniche	3	2	3	3	3
Presov	2	2.5	1	2	1
Reggio Emilia	3	3	2	3	1
Rome	2	2	2	3	2
Rotterdam	3	2	3	2	1
San Tammaro	2	3	3	3	0
Sarantaporo	3	1	3	1	3
Sassari	2	1	2	3	2
Turin	3	2	3	2	2
Utrecht	3	3	2	2	0
Wien	3	2	3	2	0
Valencia	2	2	2.5	2	1.5
Venice	2	3	2	3	1
Viladecans	2	3	3	3	2
Villeurbanne	1.5	2	2.5	2	0
Zaragoza	3	3	3	2	0

NORTH AMERICA

In the region (North America) the dataset includes twenty-three cities with eighty-one projects/public policies and closely analyzed fifteen cities and forty-one projects/public policies. US and Canadian cities received, overall, high scores and results across a number of dimensions, especially with regard to Pooling (2.6), Experimentalism (2.5), and Urban Cogovernance (2.5). As with the European cases, Tech Justice has an average score below the other dimensions (1.3). However, there are outliers (in New York City, for example) of projects pioneering the advancement of technological, digital, and data justice in cities flagging that expansion of the database is needed for a proper assessment.

Data aggregated per city

City	Urban co-governance	Enabling state	Pooling economies	Experimentalism	Tech justice
Baltimore	3	2	3	2	1
Baton Rouge	2	2	3	3	0
Boston	3	3	3	3	1
Chicago	3	3	3	3	0
Cleveland	2	2	2	2	1
Detroit	3	3	3	3	0
Jackson	2	1	3	2	1
Madison	3	1	2	2	1
Miami	2	2	2	2	2
Montreal	2.7	3	2.3	2.7	2
New York City	2	2.2	2.5	2.2	1.6
Savannah	3	2	2	3	2
Seattle	2	3	2.5	3	2
Toronto	3	2	3	2	2
Washington, DC	2	1	3	3	3

CENTRAL AND LATIN AMERICA

In the region (Central and Latin America) the dataset includes twenty cities with forty-one projects/public policies mapped, and closely analyzed ten cities and twenty-three projects/public policies. The Latin American cities received high scores in Experimentalism (2.4) and Pooling (2.2), demonstrating the presence of a developed and lively urban innovation ecosystem. Latin America has also strong scores in Urban Co-governance (2.1), Enabling State (2.2), and Tech Justice (2), thus standing as further proof of the livelihood of projects and public policies in Latin American cities, although in some cases projects/public policies analyzed are not active in the long term.

Data aggregated per city

City	Urban co-governance	Enabling state	Pooling economies	Experimentalism	Tech justice
Buenos Aires	2	1	3	3	0.5
Cochabamba	1	2	1	1	3
Medellin	1.5	2	2	1.5	2
Mexico City	1.5	3	2	3	3
Quito	2	3	3	3	2
San Josè	3	3	2	3	3
San Juan	3	2	3	3	1
Santiago de Chile	2	3	2	3	3
Sao Paolo	1	1	2	1	2
Valparaiso	3	2	3	2	1

AFRICA

In the region (Africa) the dataset includes twenty-four cities with thirty-five projects/public policies, and investigated and/or closely analyzed thirteen cities and twenty-nine projects/public policies. The case studies analyzed on the African continent are characterized by a high level of Experimentalism (2.38) and Urban Co-Governance (2.31) and a more moderate score for Pooling (2.08). African cities have the potential, in our view, to become breeding grounds for urban experimentalism and social innovation initiatives. On the other hand, cities have low scores on the dimension of the Enabling State (1.58) signaling that expansion of the database is needed for a proper assessment.

Data aggregated per city

City	Urban co-governance	Enabling state	Pooling economies	Experimentalism	Tech justice
Accra	1	2	1	3	1
Bamako	3	1	2	3	1

(*continued*)

Data aggregated per city (continued)

City	Urban co-governance	Enabling state	Pooling economies	Experimentalism	Tech justice
Bergrivier	3	3	2	2	2
Cape Town	3	2	2	2	2
Casablanca	2	1	2	3	2
Dakar	1	1	2	2	1
Johannesburg	2	2	3	3	2
Kigali	3	2	1	2	2
Kinshasa	3	1	3	3	1
Lagos	3	1	2	3	0
Lomé	2	2	3	2	3
Mombasa	2	1	2	1	1
Nairobi	3	2.5	2.5	2	0

ASIA

In the region (Asia) the dataset includes thirty-seven cities with forty-eight projects/public policies and closely analyzed twenty-five cities and thirty-six projects/public policies. Cities reported a high average score (2.2) across all the dimensons. They present a relatively moderate score for Pooling (2.2) and Experimentalism (2.2), showing noteworthy results in one of the most populated areas of the world. Tech Justice (1.4) and Enabling State (1.8) perform slightly below the average (although with notable exeptions) meaning that in these areas we may foresee scope for evolution for some of the cities included in the table below as well as the overall dataset.

Data aggregated per city

City	Urban co-governance	Enabling state	Pooling economies	Experimentalism	Tech justice
Ashdod	2	2	1	2	1
Bandung	2	2	2	3	3
Bangalore	1.5	1.5	1.5	2	2

Data aggregated per city (continued)

City	Urban co-governance	Enabling state	Pooling economies	Experimentalism	Tech justice
Banjarmasin	3	3	3	3	1
Barangay	2	2	2	3	1
Beirut	2	1	2.5	2.5	1
Chengdu	2	3	2	1.5	1
Guangzhou	2	2	2.5	2	1
Holon	3	2	1	3	1
Hong Kong	2	2	2	2	1
Jerusalem	2	2	2	3	1
Karachi	2	1	1	3	2
Kathamandu	3	1	3	3	1
Koregaoni	3	2	2	1	1
Kyoto	2	1	3	2	1
Lahore (area)	1	1	2	3	3
Mumbai	3	2	2	2	3
Pune	3	3	2	1	1
Seoul	1	2	2	2	2
Shenyang	2	2	2	1	1
Shenzhen	2	1.5	2.5	2	1
Tokyo (area)	2	1	3	1.5	3
Yogiakarta	1	2	3	2	1

OCEANIA

In the region (Oceania) the dataset includes six cities with eleven projects/public policies and closely analyzed five cities and seven projects/public policies. The following table presents the aggregated results for the region and the aggregated results per city. In Oceania, the case studies scored highly on the dimension of Pooling (2.40). They scored more moderately on Experimentalism (2.00) and lower on Urban Cogovernance (1.60), Enabling State (1.20), and Tech Justice (1.40).

Data aggregated per city

City	Urban co-governance	Enabling state	Pooling economies	Experimentalism	Tech justice
Adelaide	1	1	2	1	1
Christchurch	2	2	1	2	2
Melbourne	2	1	3	2	2
Sidney	1	1	3	2	1
Wellington	2	1	3	3	1

EXEMPLARY CASE STUDIES

Among the entire universe of collected and analyzed case studies in the first phase of this investigation, we identified the following exemplary case studies. These cases are, by no means, the only best practices our of the dataset. These are the case studies that best demonstrate and illustrate the Co-Cities principles, although they are very diverse in how they do so. They are discussed throughout the book. Our goal in the phase of the investigation that will follow the publication of this book is to identify a set of projects/public policies that is representative of different systemic variables. The case studies are the following:

City	Project/policy	Country
Amsterdam	Amsterdam Sharing City	Netherlands
Athens	SynAthina	Greece
Barcelona	Citizen Asset Regulation and Community Balance	Spain
Barcelona	Decidim	Spain
Baton Rouge	Build Baton Rouge/Co-City Baton Rouge	USA
Bangalore	Urban commons institutions for the urban lakes in Bengaluru	India
Bologna	Regulation on the collaboration between city residents and the city in the care and regeneration of the urban commons and Co-Bologna	Italy
Bologna	Iperbole Community institutional platform for the commons	Italy
Bologna	Incredibol: Bologna's creative innovation	Italy

City	Project/policy	Country
Bologna	Co-Bologna (fieldwork and experimentation)	Italy
Boston	Dudley Street	USA
Lomé	Woelab	Togo
Madrid	Ordinance on public social cooperation	Spain
Milan	Deliberation on the criteria for use and concession of use of city-owned buildings for projects aimed at social, cultural, economic development.	Italy
Mexico City	Laboratorio para la ciudad	Mexico
Mexico City	Ciudad Propuesta CDMX—Proposed City CDMX	Mexico
Naples	Deputy Mayor for the Commons	Italy
Naples	Agency for the Water as a Commons (ABC Naples)	Italy
Naples	Principles for the governance of the urban commons and Urban Civic Uses Recognition	Italy
Naples	Ex Asilo Filangieri	Italy
Naples	Civic eState URBACT transfer network (Co-City experimentation)	Italy
New York City	MOCTO/NYCx Co-Lab	USA
New York City	Silicon Harlem	USA
New York City	Red Hook Wi-Fi	USA
Reggio Emilia	Neighborhood as a commons	Italy
Reggio Emilia	Coviolo Wireless	Italy
Reggio Emilia	Collaboratorio Reggio (fieldwork and experimentation)	Italy
Rome	Agenda Tevere	Italy
Rome	Co-Roma social partnership (fieldwork and experimentation)	Italy
San Juan	Community Land Trust	Puerto Rico
Seoul	Municipal Ordinance for the Sharing Economy	Korea
Turin	Co-City Torino	Italy
Turin	New Turin Regulation for Governing the Urban Commons	Italy
Turin	Neighborhood houses network	Italy
Turin	Co-City (UIA project) (fieldwork)	Italy

NOTES

3. THE CITY AS A COMMONS

1. Seoul Metropolitan Government Ordinance on the Promotion of Sharing Enactment § 5396 (passed Dec. 31, 2012). https://legal.seoul.go.kr/legal/english/front/page/law.html?pAct=lawView&pPromNo=1191.

2. Seoul Innovation Bureau (City Transition Division) Ecological Inclusive City based on civil autonomy, The 3rd Sharing City Seoul Master Plan (2012–2025).

3. Seoul Innovation Bureau (City Transition Division) Ecological Inclusive City based on civil autonomy, The 3rd Sharing City Seoul Master Plan (2012–2025), 12–13.

4. Misure urgenti per il sostegno a famiglie, lavoro, occupazione e impresa e per ridisegnare in funzione anti-crisi il quadro strategico nazionale, Gov't Decree Legis. 185 (Passed November 19, 2008), art 23. https://www.parlamento.it/parlam/leggi/decreti/08185d.htm.

5. City of Bologna, Promotion of Active Citizenship, Report attività 2014-2016, http://partecipa.comune.bologna.it/sites/comunita/files/allegati_blog/due_anni_di_patti_di_collaborazione_relazione_2014-2016.pdf.

6. City of Bologna, Rendicontazione sociale rapporti con il Terzo Settore e Cittadinanza Attiva, http://partecipa.comune.bologna.it/rendicontazione-sociale-rapporti-con-il-terzo-settore-e-cittadinanza-attiva.

7. Naples City Council, Resolution n. 7, Guidelines for the Identification and Management of City-Owned Buildings, Dismissed or Partially Used, That Are Perceived by the Community as Commons and Are Susceptible to a Form of Collective Use (Passed March 9, 2015). https://www.comune.napoli.it/flex/cm/pages/ServeBLOB.php/L/IT/IDPagina/16783#:~:text=Nel%202012%20%C3%A8%20stato%20approvato,ogni%20cittadino%20deve%20concorrere%20al: Naples City Government,

Resolution n. 446, Recognition Pursuant to the City Council Resolution n. 7/2015. Identification of Spaces of Civic Relevance to be Recognized as Commons (Passed May 26, 2016). https://www.comune.napoli.it/flex/cm/pages/ServeAttachment.php/L/IT/D/5%252F5%252F6%252FD.f3d51671fdbfa028027c/P/BLOB%3AID%3D16783/E/pdf.

REFERENCES

Ackerman, 2004. Co-Governance for Accountability: Beyond "Exit" and "Voice." *World Development* 32, no. 3 (March): 447–463. https://doi.org/10.1016/j.worlddev.2003.06.015.

Acuto, Michele. 2018. Global Science for City Policy, *Science* 359, no. 6372 (January): 165–166.

Adler, David. 2017. The Fragmented City: Mexico City and the Right to the City Charter. *The Right to the City: A Verso Report*, 90–98. London: Veso.

Agyeman, Julian, Robert Bullard, and Bob Evans, eds. 2003. *Just Sustainabilities: Development in an Unequal World*. Cambridge MA: MIT Press.

Ayuntamiento de Madrid. 2018a. Ordenanza de Cooperación Público—Social del Ayuntamiento de Madrid. núm. 8173 pág. 4–24. https://sede.madrid.es/portal/site/tramites/menuitem.5dd4485239c96e10f7a72106a8a409a0/?vgnextoid=eb53ebd8dfcf3610VgnVCM2000001f4a900aRCRD&vgnextchannel=6b3d814231ede410VgnVCM1000000b205a0aRCRD&vgnextfmt=default.

Ayuntamiento de Madrid. Área de Gobierno de Vicealcaldía. 2020a. Memoria propuesta para el trámite de la consulta pública previa sobre la propuesta de derogación de la ordenanza de cooperación público-social, para su integración parcial en el reglamento orgánico de participación ciudadana del ayuntamiento de madrid, así como sobre la modificación del mismo. *Transparencia.madrid.es* (website). https://transparencia.madrid.es/FWProjects/transparencia/InformacionJuridica/HuellaNormativa/Organicos/ROParticipacion2020/Ficheros/20200119MemoriaConsultaPublica.pdf.

Ayuntamiento de Madrid. Área de Gobierno de Vicealcaldía. 2020b. Consulta pública previa a la derogación de la ordenanza de cooperación público social del ayuntamiento de madrid, para su integración parcial en el reglamento orgánico de participación ciudadana del ayuntamiento de madrid, así como la modificación del mismo. *Transparencia.madrid.es* (website). https://transparencia.madrid.es/FWProjects/transparencia/InformacionJuridica/HuellaNormativa/Organicos/ROParticipacion2020/Ficheros/20200205InformeResultadoConsulta.pdf.

Albanese, Rocco, and Elisa Michelazzo. 2020. *Manuale di Diritto dei Beni Comuni Urbani*. Torino: Celid.

Albouy, David, Gabriel Ehrlich, and Minchul Shin. 2018. Metropolitan Land Values. *Review of Economics and Statistics* 100, no. 3 (July): 454–466.

Alexander, Gregory S. 2009. The Social-Obligation Norm in American Property Law. *Cornell Law Review* 94, no. 4 (May): 745–820.

Alexander, Gregory S. 2012. Governance Property. *University of Pennsylvania Law Review* 160, no. 7 (June): 1853–1887.

Alexander, Gregory S. 2020. The Human Flourishing Theory. *Cornell Legal Studies Research Paper* 20, no. 2 (February).

Alexander, Gregory S., and Eduardo M. Peñalver. 2009. Properties of Communities. *Theoretical Inquiries in Law* 10, no. 1 (January): 127–160.

Alexander, Lisa T. 2015. Occupying the Constitutional Right to Housing. *Nebraska Law Review* 94, no. 2 (October): 245–301.

Alexander, Lisa T. 2019. Community in Property: Lessons from Tiny Homes Villages. *Minnesota Law Review* 104, no. 1 (October): 385–402.

Algoed, Line, María E. Hernández Torrales, and Lyvia Rodríguez Del Valle. 2018. El Fideicomiso de la Tierra del Caño Martín Peña Instrumento Notable de Regularización de Suelo en Asentamientos Informales. *Lincoln Institute of Land Policy, Working paper* 18, no. 1. https://www.lincolninst.edu/sites/default/files/pubfiles/algoed_wp18la1sp.pdf.

Aligica, Paul Dragos, and Vlad Tarko. 2012. Polycentricity: From Polanyi to Ostrom, and Beyond. *Governance* 25, no. 2 (September): 237–262.

Alvarez, Sarah, and Leah Samuel. 2018. Real Estate is Hot in Detroit. But its Top Owner, the City, Isn't Selling. *Bridge News*, August 21, 2018. https://www.bridgemi.com/detroit-journalism-cooperative/real-estate-hot-detroit-its-top-owner-city-isnt-selling.

Ambole, Amollo, Kweku Koranteng, Peris Njoroge, and Douglas Logedi Luhangala. 2021. A Review of Energy Communities in Sub-Saharan Africa as a Transition Pathway to Energy Democracy. *Sustainability* 13, no. 4: 2128.

Anderson, Michelle Wilde. 2014. The New Minimal Cities. *Yale Law Journal* 123, no. 5 (March): 1118–1227.

Ansell, Chris, and Alison Gash. 2008. Collaborative Governance in Theory and Practice. *Journal of Public Administration Research and Theory* 18, no. 4 (October): 543–571. https://doi.org/10.1093/jopart/mum032.

Antonelli, Alessandro, Elena De Nictolis, and Christian Iaione. 2021. QUA. Quartiere Bene Comune. *Urban Maestro* (website). Accessed February 20, 2022. https://urbanmaestro.org/example/quartiere-bene-comune/.

Araral, Eduardo, and Kris Hartley. 2013. Polycentric Governance for a New Environmental Regime: Theoretical Frontiers in Policy Reform and Public Administration. In *Proceedings of the First International Conference on Public Policy, Grenoble, France, June 26–28*. Bloomington: Indiana University.

Arnstein, Sherry. 1969. A Ladder of Citizen Participation. *Journal of the American Planning Association* 35, no. 4: 216–244.

Attoh, Kafui. 2011. What Kind of Right Is the Right to the City? *Progress in Human Geography*. 35, no. 5 (February): 669–685.

Atuahene, Bernadette. 2020. Predatory Cities. *California Law Review* 108, no. 1 (February): 107–182.

Baccarne, Bastiaan, Dimitri Schuurman, Peter Mechant, and Lieven De Marez. 2014. The Role of Urban Livng Labs in a Smart City. *Technology Innovation Management Review* 6, no. 3: 22–30.

Baer, Susan E., and Richard C. Feiock. 2005. Private Governments in Urban Areas: Political Contracting and Collective Action. *American Review of Public Administration* 35, no. 1 (March): 42–56. https://doi.org/10.1177/0275074004271717.

Bailey, Saki, and Maria Edgarda Marcucci. 2013. Legalizing the Occupation: The Teatro Valley as a Cultural Commons. *South Atlantic Quarterly* 112, no. 2 (April): 396–405. https://doi.org/10.1215/00382876-2020271.

Bailey, Saki, and Ugo Mattei. 2013. Social Movements as Constituent Power: The Italian Struggle for the Commons. *Indiana Journal of Global Legal Studies* 20, no. 2 (Summer): 965–1013. https://www.repository.law.indiana.edu/ijgls/vol20/iss2/14.

Baiocchi, Gianpaolo. 2003. *Radicals in Power: The Workers' Party Pt and Experiments in Urban Democracy in Brazil*. London: Zed Books.

Baiocchi, Gianpaolo, and Ernesto Ganuza. 2014. Participatory Budgeting As If Emancipation Mattered. *Politics and Society* 42, no. 1 (January): 29–50. https://doi.org/10.1177/0032329213512978.

Baker, Rachael. 2020. Racial Capitalism and a Tentative Commons. In *Commoning the City: Empirical Perspectives on Urban Ecology, Economics, and Ethics*, edited by Derya Ozkan and Guldem Baykal Buyuksara, 25–36. Oxfordshire, UK: Routledge, Taylor & Francis.

Bang, Henrik. 2010. Between Everyday Makers and Expert Citizens. In *Public Management in the Postmodern Era: Challenges and Prospects*, edited by John Fenwick and Janice McMillan, 163–92. Cheltenham, UK: Edward Elgar Publishing.

Barca, Fabrizio. 2009. *An Agenda For A Reformed Cohesion Policy: A Place-based Approach to Meeting European Union Challenges and Expectations*. Independent Report prepared at the request of Danuta Hübner, Commissioner for Regional Policy. European Commission (website). Accessed March 15, 2022. https://ec.europa.eu/migrant-integration/library-document/agenda-reformed-cohesion-policy-place-based-approach-meeting-european-union_en.

Barcelona City Council. 2017a. *Modificación Pla General Metropolitano en el Ambito de Can Batlló-Magòria, Ayuntament de Barcelona*. Online Geographic Information system of the Barcelona City Council (website). Accessed September 10, 2021. https://ajuntament.barcelona.cat/informaciourbanistica/cerca/es/fitxa/B1524/--/--/ap/.

Barcelona City Council. 2017b. *Reglament de Participació, Ajuntament de Barcelona Ciutadana*. Barcelona City Council. https://ajuntament.barcelona.cat/participaciociutadana/sites/default/files/documents/reglament_participacio_catala.pdf.

Barcelona City Council. 2019a. *Barcelona City Council Commons Policy: Citizens Asset Programme and Community Management of Public Resources and Services*. Barcelona Citizen Participation—Area of Culture, Education, Science, and Education. https://ajuntament.barcelona.cat/participaciociutadana/sites/default/files/documents/barcelona_city_council_commons_policy_citizen_assets_programme.pdf.

Barcelona City Council. 2019b. *Citizen Property*. Barcelona Citizen Participation—Area of Culture, Education, Science, and Education (Power Point presentation). https://ajuntament.barcelona.cat/participaciociutadana/sites/default/files/documents/barcelona_city_council_commons_policy_power_point.pdf.

Barritt, Emily. 2020. The Story of Stewardship and Ecological Restoration. In *Ecological Restoration Law: Concepts and Case Studies*, edited by Afshin Akhtar-Khavari and Benjamin J. Richardson. New York: Routledge.

Bauwens, Michel, and Pantazis Alekos. 2018. The Ecosystem of Commons-Based Peer Production and Its Transformative Dynamics. *Sociological Review* 66, no. 2 (March): 302–319.

Bauwens, Michael, and Yurek Onzia. 2017. *Commons Transition Plan for the City of Ghent*. Ghent Municipality (public document). https://stad.gent/sites/default/files/page/documents/Commons%20Transition%20Plan%20-%20under%20revision.pdf.

Belli, Luca. 2016. *Community Connectivity: Building the Internet from Scratch. Annual Report of the UN IGF Dynamic Coalition on Community Connectivity*. Internet Governance Forum (IGF). https://bibliotecadigital.fgv.br/dspace/handle/10438/17528.

Benkler, Yochai. 2004. Sharing Nicely: On Sharable Goods and the Emergence of Sharing as a Modality of Economic Production. *Yale Law Journal* 114, no. 2 (November): 273–358.

Benkler, Yochai. 2006. *The Wealth of Networks: How Social Production Transforms Markets and Freedom*. New Haven, CT: Yale University Press.

Benkler, Yochai. 2016. Open Access and Information Commons. In *Oxford Handbook of Law and Economics: Private and Commercial Law*, edited by Francesco Parisi. Oxford: Oxford University Press.

Benkler, Yochai, and Helen Nissenbaum. 2006. Commons-Based Peer Production and Virtue. *Journal of Political Philosophy* 14, no. 4: 394–419. https://ssrn.com/abstract=2567434.

Bennett, Nathan J., Tara S. Whitty, Elena Finkbeiner, Jeremy Pittman, Hannah Bassett, Stefan Gelcich, and Edward H. Allison. 2018. Environmental Stewardship: A Conceptual Review and Analytical Framework. *Environmental Management* 61 (January): 597–614. https://doi.org/10.1007/s00267-017-0993-2.

Bentley, George C., Priscilla McCutcheon, Robert G. Cromley and Dean M. Hanink. 2016. Race, Class, Unemployment, and Housing Vacancies in Detroit: An Empirical Analysis. *Urban Geography* 37, no. 5: 785–800. https://doi.org/10.1080/02723638.2015.1112642.

Bernardi, Monica. 2017a. An Informal Settlement as a Community Land Trust. The case of San Juan, Puerto Rico. *LabGov.org* (blog). July 20, 2020. http://labgov.city/thecommonspost/the-favela-as-a-community-land-trust-the-case-of-san-juan-puerto-rico.

Bernardi, Monica. 2017b. Sharing and cooperative practices for local sustainable development: the urban village communities in Seoul. The case of the Sungmisan Village. *LabGov.city*, https://labgov.city/theurbanmedialab/sharing-and-cooperative-practices-for-local-sustainable-development-the-urban-village-communities-in-seoul-the-case-of-the-sungmisan-village/.

Bernardi, Monica, and Davide Diamantini. 2018. Shaping the Sharing City: An Exploratory Study on Seoul and Milan. *Journal of Cleaner Production*. 203 (December): 30–42. https://doi.org/10.1016/j.jclepro.2018.08.132.

Bertello, Agnese. 2012. L'esperienza delle Case del Quartiere di Torino. *ItalianiEuropei* 1–5: 183–188. https://casedelquartiere.files.wordpress.com/2013/01/lesperienza-delle-case-di-quartiere-a-torino_italianieuropei.pdf.

Berti Suman, Anna, and Marina van Geenhuizen. 2020. Not Just Noise Monitoring: Rethinking Citizen Sensing for Risk-Related Problem-Solving. *Journal of Environmental Planning and Management* 63, no. 3 (April): 546–567.

Berti Suman, Anna, and Robin Pierce. 2018. Challenges for Citizen Science and the EU Open Science Agenda under the GDPR. *European Data Protection Law Review* 4, no 3: 284–295.

Bianchi, Iolanda. 2018. The Post-Political Meaning of the Concept of Commons: The Regulation of the Urban Commons in Bologna. *Space and Policy* 22, no. 3 (August): 287–306. https://doi.org/10.1080/13562576.2018.1505492.

Bina, Olivia, Josefine Fokdal, Luís Balula, and Marta Pedro Varanda. 2015. Getting the Education for the City We Want. In *Urban Pamphleeter no. 5. Global Education for Urban Futures*, edited by Paola A. d'Alençon, Ben Campkin, Rupali Gupte, Mkhabela Solam, Johannes Novy, and Mika Savela, 7–9. London: UCL Urban Laboratory. http://hdl.handle.net/10451/20279.

Bingham, Lisa. 2009. Collaborative Governance: Emerging Practices and the Incomplete Legal Framework for Public and Stakeholder Voice. *Journal of Dispute Resolution* 2009, no. 2 (July): 270–291. https://scholarship.law.missouri.edu/jdr/vol2009/iss2/2.

Bingham, Lisa Blomgren. 2010. The Next Generation of Administrative Law: Building the Legal Infrastructure for Collaborative Governance. *Wisconsin Law Review* 297 (July): 297–356.

Blanco, Ismael. 2009. Does a "Barcelona model" Really Exist? Periods, Territories and Actors in the Process of Urban Transformation. *Local Government Studies* 35, no. 3 (July): 355–369. https://doi.org/10.1080/03003930902854289.

Blanco, Ismael. 2015. Between Democratic Network Governance and Neoliberalism: A Regime-Theoretical Analysis of Collaboration in Barcelona. *Cities* 44 (April): 123–130. https://doi.org/10.1016/j.cities.2014.10.007.

Blanco, Ismael, Yunailis Salazar, and Iolanda Bianchi. 2020. Urban Governance and Political Change Under a Radical Left Government: The Case of Barcelona. *Journal of Urban Affairs* 42, no. 1: 18–38. https://doi.org/10.1080/07352166.2018.1559648.

Blomley, Nicholas. 2008. Enclosure, Common Right, and the Property of the Poor. *Social Legal Studies* 17, no. 3 (September): 311–331.

Boggs, Grace Lee. 2012. Detroit, Place, and Space to Begin Anew. In *The Next American Revolution: Sustainable Activism for the Twenty-First Century*. Berkeley: University of California Press.

Bollier, David, and Silke Helfrich. eds. 2015. *Patterns of Commoning*. Amherst, MA: The Commons Strategies Group in cooperation with Off the Common Books.

Borch, Christian, and Martin Kornberger. 2015. Introduction. In *Urban Commons: Rethinking the City*. Oxfordshire, UK: Routledge.

Botsman, Rachel, and Roo Rogers. 2010. *What's Mine Is Yours*. New York: HarperCollins.

Bresnihan, Patrick, and Michael Byrne. 2015. Escape into the City: Everyday Practices of Commoning and the Production of Urban Space in Dublin. *Antipode* 47, no. 1 (July): 36–54. https://doi.org/10.1111/anti.12105.

Breznitz, Shiri M., and Maryann P. Feldman. 2012. The Engaged University. *Journal of Technology Transfer* 37 (July):139–157.

Briffault, Richard. 1999. A Government for Our Time? Business Improvement Districts and Urban Governance. *Columbia Law Review* 99: 365–477. https://doi.org/10.2307/1123583.

Bugg-Levine, Anthony, and Jed Emerson. 2011. Impact investing: Transforming How We Make Money While Making a Difference. *The MIT Press Journal* 6, no. 3: 9–18. https://EconPapers.repec.org/RePEc:tpr:inntgg:v:6:y:2011:i:3:p:9-18.

Bunce, Susannah. 2015. Pursuing Urban Commons: Politics and Alliances in Community Land Trust Activism in East London. *Antipode* 48, no. 1 (June): 134–150. https://doi.org/10.1111/anti.12168.

Calafati, Antonio G., and Paolo Veneri. 2013. Re-defining the Boundaries of Major Italian Cities. *Regional Studies* 47, no. 5 (August): 789–802. https://doi.org/10.1080/00343404.2011.587798.

Caldeira, Teresa P. R. 2017. Peripheral Urbanization: Autoconstruction, Transversal Logics, and Politics in Cities of the Global South. *Environment and Planning D: Society and Space* 35, no. 1 (February): 3–20. https://doi.org/10.1177/0263775816658479.

Campbell, Lindsay K., Erika Svendsen Johnson, Michelle Johnson, and Laura Landau. 2021. Activating Urban Environments as Social Infrastructure through Civic Stewardship. *Urban Geography* (May):1–22. https://doi.org/10.1080/02723638.2021.1920129.

Caponio, Tiziana, and Davide Donatiello. 2017. Intercultural Policy in Times of Crisis: Theory and Practice in the Case of Turin, Italy, *Comparative Migration Studies* 5, 13. https://doi.org/10.1186/s40878-017-0055-1.

Carayannis, Elias, Thorsten D. Barth, and David F. J. Campbell. 2012. The Quintuple Helix Innovation Model: Global Warming as a Challenge and Driver for Innovation. *Journal of Innovation and Entrepreneurship* 1, no. 2 (August).

Carayannis, Elias, and David F. J. Campbell. 2009. "Mode 3" and "Quadruple Helix": Toward a 21st Century Fractal Innovation Ecosystem. *International Journal of Technology Management* 46, no. 3/4 (January): 201–234.

Carlisle, Keith, and Rebecca L. Gruby. 2017. Polycentric Systems of Governance: A Theoretical Model for the Commons. *Policy Studies Journal* 47, no. 4 (August): 927–952. https://doi.org/10.1111/psj.12212.

Carra, Martina, Nicoletta Levi, Giulia Sgarbi, and Chiara Testoni. 2018. From Community Participation to Co-design: "Quartiere bene comune" Case Study. *Journal of Place Management and Development* 11, no. 2 (May): 242–258.

Cassese, Sabino. 2017. From the Nation-State to the Global Polity. In *Reconfiguring European States*, edited by Desmond King and Patrick Les Galès, 78–96. Oxford: Oxford University Press.

Castro, Mauro, Iolandra Fresnillo, and Rubén Martínez. 2016. *Comuns urbans—Patrimoni Ciutadà. Marc conceptual i propostes de línies d'acció*. Barcelona City Council, Directorate of Active Democracy. http://lahidra.net/wp-content/uploads/2017/07/Patrimoni-Ciutada-Marc-Conceptual-v.3.01.pdf.

Celauro, A., M. Marsella, P. J. V. D'Aranno, A. Maass, J. A. Palenzuela Baena, J. F. Guerrero Tello, and I. Moriero. 2019. Ancient Mining Landscapes and Habitative Sceneries in the Urban Area of Centocelle: Geomatic Applications for their Identification, Measurement, Documentation and Monitoring. *International Archives of the Photogrammetry, Remote Sensing and Spatial Information Sciences* XLII-2/W1 (May): 403–410. 10.5194/isprs-archives-XLII-2-W11-403-2019.

Cellamare, Carlo. 2017. Epiphanic Peripheries, Re-appropriation of the City and Dwelling Quality. *Quolibet* U3:14; 2:53–62. https://www.torrossa.com/it/resources/an/4466659.

Cepiku, Denita, Domenico Ferrari, and Angela Greco. 2006. Governance e coordinamento strategico delle reti di aziende sanitarie. *Mecosan: management ed economia sanitaria* 57: 17–36.

Charnock, Greig, Hug March, and Ramon Ribera-Fumaz. 2019. From Smart to Rebel City? Worlding, Provincialising and the Barcelona Model. *Urban Studies* 58, no. 3 (October): 581–600.

Chatterton, Paul, Alice Owen, Jo Cutter, Gary Dymski, and Rachael Unsworth. 2018. Recasting Urban Governance through Leeds City Lab: Developing Alternatives to Neoliberal Urban Austerity in Co-production Laboratories. *International Journal of Urban and Regional Research* 42, no. 2 (March): 226–243. https://doi.org/10.1111/1468-2427.12607.

Cho, Michelle. 2016. Benefit Corporations in the United States and Community Interest Companies in the United Kingdom: Does Social Enterprise Actually Work, *Northwestern Journal of International Law & Business* 37: 149. http://scholarlycommons.law.northwestern.edu/njilb?utm_source=scholarlycommons.law.northwestern.edu%2Fnjilb%2Fvol37%2Fiss1%2F4&utm_medium=PDF&utm_campaign=PDFCoverPages.

Chronéer, Diana, Anna Ståhlbröst, and Habibipour Abdolrasoul. 2019. Urban Living Labs: Towards an Integrated Understanding of their Key Components. *Technology Innovation Management Review* 9, no. 3 (March): 50–62.

Chung, Him, and Jonathan Unger. 2013. The Guangdong Model of Urbanization: Collective Village Land and the Making of a New Middle Class. *China Perspectives* 2013 (3): 33–41. https://doi.org/10.4000/chinaperspectives.6258.

CIC Regulator. n.d. UK Government (website). Accessed March 13, 2022. https://www.gov.uk/government/organisations/office-of-the-regulator-of-community-interest-companies.

Cirillo, Lidia. 2014. *Lotta di classe sul palcoscenico: i teatri occupati si raccontano*. Rome, Italy: Alegre.City of Amsterdam. 2019. *Room for Initiative*. City of Amsterdam (website). Accessed August 19, 2020. https://www.amsterdam.nl/veelgevraagd/?productid=%7B1D452E36-8F3E-46F4-9854-BF98FF0C5AAA%7D.

City of Seattle, Seattle Department of Neighborhoods. 2020. City of Seattle Will Transfer Fire Station 6 to Community. *Front Porch*, June 12, 2020. https://frontporch.seattle.gov/2020/06/12/city-of-seattle-will-transfer-fire-station-6-to-community/.

Civic eState. 2020. *Legal Principles and Tools for Urban Commons Governance*. Civic eState (website). Accessed January 24, 2022. http://www.civicestate.eu/barcelona/.

Civic eState. 2021a. *Civic Estate—Pooling the Urban Commons*. Civic eState (website). Accessed June 21, 2021. https://www.civicestate.eu/.

Civic eState. 2021b. *CIVIC eSTATE: New Models of Urban Co-governance Based on the Commons. Civic eState Final Network Product*. Civic eState (website). July 2021. http://www

.civicestate.eu/wp-content/uploads/2021/11/Final-product-CIVIC-eSTATE-new-models-of-urban-co-governance-based-on-the-commons-1.pdf.

Co-Bologna. n.d.a. *Restituzione finale del Cantiere Bolognina*. LABoratorio per la GOVernance dei Beni Comuni di Bologna (website). Accessed September 10, 2021. https://co-bologna.it/2016/11/15/restituzione-finale-del-cantiere-bolognina/.

Co-Bologna. n.d.b. *Restituzione finale del Cantiere Pilastro*. LABoratorio per la GOVernance dei Beni Comuni di Bologna (website). Accessed March 20, 2021.

Co-Bologna. n.d.c. *Restituzione finale del Cantiere Piazza dei Colori*. LABoratorio per la GOVernance dei Beni Comuni di Bologna (website). Accessed March 20, 2021. https://co-bologna.it/2016/10/22/restituzione-finale-del-cantiere-piazza-dei-colori/.

Colding, Johan, Stephan Barthel, Pim Bendt, Robbert Snep, Wim Van der Knaap, and Henrik Ernstson. 2013. Urban Green Commons: Insights on Urban Common Property Systems. *Global Environmental Change* 23, no. 5 (October): 1039–1051. https://doi.org/10.1016/j.gloenvcha.2013.05.006.

Cole, Daniel H. 2011. From Global to Polycentric Climate Governance. *Climate Law* 2: 395–413. https://doi.org/10.3233/CL-2011-042.

Cole, Luke W., and Sheila R. Foster. 2001. *From the Ground Up: Environmental Racism and the Rise of the Environmental Justice Movement*. New York: NYU Press.

Comité National de Liaison des Régies de Quartier (CNLRQ). 2017. *Régies de Quartier et Régies de Territoire*. Comité National de Liaison des Régies de Quartier. https://www.regiedequartier.org/wp-content/uploads/2020/05/2020-05-CarteRegiesQT-vf-1.pdf.

Committee of the Regions. 2015. Opinion of the European Committee of the Regions—The Local and Regional Dimension of the Sharing Economy. (2016/C 051/06).

Comune di Bologna. n.d. *Bologna e I Beni Comuni Urbani*. Iperbole, ReteCivica (website). http://partecipa.comune.bologna.it/beni-comuni.

Comune di Roma. n.d. *Casa Delle Tecnologie Emergenti: L'ecosistema Dei Talenti Dell'innovazione Digitale*. ComuneRoma (website). https://www.comune.roma.it/eventi/it/roma-innovation-progetto-smartcity.page?contentId=PRG38591.

Cook, Graeme, and Rick Muir. 2012. The Relational State. *Institute for Public Policy Research*. https://www.ippr.org/files/images/media/files/publication/2012/11/relational-state_Nov2012_9888.pdf.

Cossetta, Anna, and Mauro Palumbo. 2014. The Co-production of Social Innovation Social Innovation: The Case of Living Lab. In Dameri, R., Rosenthal-Sabroux, C. (eds). *Smart City. Progress in IS*. Cham, Switzerland: Springer. https://doi.org/10.1007/978-3-319-06160-3_11.

Cruz Ruiz, E., 2021. Underrepresented Groups and Constitution-Making: The Mexico City Case. *Political Studies Review* 19, no. 2: 164–170. https://doi.org/10.1177/1478929920944825.

Cummings, Scott L. 2001. Community Economic Development as Progressive Politics: Toward a Grassroots Movement for Economic Justice. *Stanford Law Reviw* 54: 399.

Cummings, Scott L. 2006. Mobilization Lawyering: Community Economic Development in the Figueroa Corridor. In *Cause Lawyers and Social Movements*, edited by Austin Sarat and Stuart A. Scheingold. Stanford, CA: Stanford University Press.

da Cruz, Nuno F., Philipp Rode, and Michael McQuarrie. 2018. New Urban Governance: A Review of Current Themes and Future Priorities. *Journal of Urban Affairs* 41, no. 2 (August): 1–19. https://doi.org/10.1080/07352166.2018.1499416.

Dahl, Robert. 1967. The City in the Future of Democracy. *American Political Science Review* 61, no. 4 (December): 953–970. https://doi.org/10.2307/1953398.

Dahlmann, Frederik, Wendy Stubbs, Rob Raven, and João Porto de Albuquerque. 2020. The "Purpose Ecosystem": Emerging Private Sector Actors in Earth System Governance. *Earth System Governance* 4, (June): 100053.

d'Albergo, Ernesto, and Daniela De Leo. 2018. *Politiche urbane per Roma Le sfide di una Capitale debole*. Rome: Sapienza University Press.

D'Alena, Michele. 2021. *Immaginazione civica: L'energia delle comunita dentro la politica*. Bologna: Luca Sossella Editore.

Daniels, Brigham. 2007. Emerging Commons and Tragic Institutions. *Environmental Law* 37, no. 3 (Summer): 515–571.

David, Cristina C., and Arlene B. Inocencio. 2001. Public-Private-Community Partnerships in Management and Delivery of Water to Urban Poor: The Case of Metro Manila. *Discussion Papers DP 2001–18, Philippine Institute for Development Studies*

Davidson, Nestor, and Sheila Foster. 2013. The Mobility Case for Regionalism. *U.C. Davis Law Review*, 47 (November): 63–120.

Davidson, Nestor M., and John J. Infranca. 2016. The Sharing Economy as an Urban Phenomenon. *Yale Law & Policy Review* 34, no. 2 (June): 215–279. https://digitalcommons.law.yale.edu/cgi/viewcontent.cgi?article=1697&context=ylpr.

Davis, J. E. 2010. Origins and Evolution of the Community Land Trust in the United States. In *The Community Land Trust Reader*, edited by J. E. Davis, 3–47. Cambridge, MA: Lincoln Institute of Land Policy.

Davis, Mike. 2006. *Planet of Slums*. London: Verso.

Deakin, Mark. 2014. Smart Cities: The State of the Art and Governance Challenge. *Triple Helix* 1, no. 7 (November): 1–16. https://doi.org/10.1186/s40604-014-0007-9.

De Angelis, Massimo. 2017. *Omnia Sunt Communia: On the Commons and the Transformation to Postcapitalism*. London: Zed Books.

De Barbieri, Edward. 2016. Do Community Benefits Agreements Benefit Communities? *Cardozo Law Review* 37, no. 5 (June): 1773–1825.

De Búrca, Gráinne, Robert O. Keohane, and Charles F. Sabel. 2014. Global Experimentalist Governance. *British Journal of Political Science* 44, no. 3: 477–486.

De Filippi, Primavera, and Félix Tréguer. 2015. Wireless Community Networks: Towards a Public Policy for the Network Commons? In *Net Neutrality Compendium*, edited by Luca Belli and Primavera De Filippi. Cham, Switzerland: Springer.

De Filippis, James, Brian Stromberg, and Olivia R. Williams. 2018. W(h)ither the Community in Community Land Trusts? *Journal of Urban Affairs* 40, no. 6: 755–769. https://doi.org/10.1080/07352166.2017.1361302.

Dellenbaugh, Mary, Agnes K. Muller, Markus Kip, Majken Bieniok, and Martin Schwegmann, eds. 2015. Urban Commons: Moving Beyond State and Market. *Bauwelt Fundamente* 154 (June). https://doi.org/10.1515/9783038214953.

Delsante, Ioanni, and Nadia Bertolino. 2017. Urban Spaces Commoning and Its Impact on Planning: A Case Study of the Former Slaughterhouse Exchange Building in Milan. *Der öffentliche Sektor* 43, no. 1 (June): 45–56. https://doi.org/10.34749/oes.2017.2384.

De Moor, Tine. 2012. What Do We Have in Common? A Comparative Framework for Old and New Literature on the Commons. *International Review of Social History* 57, no. 2 (August): 269–290.

Detroit Digital Justice Coalition. n.d. Principles. (website) Accessed September 10, 2021. http://detroitdjc.org/principles/.

De Tullio, Maria Francesca. 2018. Commons towards New Participatory Institutions: The Neapolitan Experience. In *Exploring Commonism—A New Aesthetics of the Real*, edited by P. Gielen N. Dockx. Amsterdam: Valiz.

De Tullio, Maria Francesca, and Roberto Cirillo, eds. 2021. *Healing Culture, Reclaiming Commons, Fostering Care: A Proposal for EU Cultural Policies*. Naples, Italy: IF press.

di Robilant, Anna. 2012. Common Ownership and Equality of Autonomy. *McGill Law Journal* 58, no. 2: 263–320. https://scholarship.law.bu.edu/cgi/viewcontent.cgi?article=1074&context=faculty_scholarship.

di Robilant, Anna. 2014. Property and Democratic Deliberation: The Numerus Clausus Principle and Democratic Experimentalism in Property Law. *American Journal of Comparative Law* 62, no. 2 (Spring): 367–416. https://ssrn.com/abstract=2456658.

Dobbs, Richard, James Manyika, Jonathan Woetzel, Jaana Remes, Jesko Perrey, Greg Kelly, Kanaka Pattabiraman, and Hemant Sharma. 2016. Urban World: The Global Consumer to Watch. *McKinsey Global Institute* (blog). March 30, 2016. https://www.mckinsey.com/featured-insights/urbanization/urban-world-the-global-consumers-to-watch#.

Dryzek, John S. 2009. Democratization as Deliberative Capacity Building. *Comparative Political Studies* 42, no. 11 (November): 1379–1402. https://doi.org/10.1177/0010414009332129.

D'Souza, R., and H. Nagendra. 2011. Changes in Public Commons as a Consequence of Urbanization: The Agara Lake in Bangalore, India. *Environmental Management* 47, no. 5 (May): 840–850. https://doi.org/10.1007/s00267-011-9658-8.

Dutz, M., Y. Kuznetsov, E. Lasagabaster, and D. Pilat, eds. 2014. *Making Innovation Policy Work: Learning from Experimentation*. Paris, France: OECD. https://www.oecd-ilibrary.org/making-innovation-policy-work_5k91gw7bzd7b.pdf.

Ehrenhalt, Alan. 2012. *The Great Inversion and the Future of the American City*. New York: Vintage.

Einstein, Katherine L., Maxwell Palmer, and David M. Glick. 2018. Who Participates in Local Government? Evidence from Meeting Minutes. *Perspectives on Politics* 17, no. 1 (October): 1–19.

Ela, Nate. 2016. Urban Commons as Property Experiment: Mapping Chicago's Farms and Gardens. *Fordham Urban Law Journal* 43, no. 2: 247–294. https://ir.lawnet.fordham.edu/ulj/vol43/iss2/2.

Ellickson, Robert C. 1991. *Order Without Law*. Cambridge, MA: Harvard University Press.

Ellickson, Robert C. 1996. Controlling Chronic Misconduct in City Spaces: Of Panhandlers, Skid Rows, and Public-Space Zoning. *Yale Law Journal* 105, no. 5 (January): 1165–1248.

Elwood, Sarah. 2002. Neighborhood Revitalization Through "Collaboration": Assessing the Implications of Neoliberal Urban Policy at the Grassroots. *GeoJournal* 58 (October): 121–130. https://doi.org/10.1023/B:GEJO.0000010831.73363.e3.

Elwood, Sarah. 2004. Partnerships and Participation: Reconfiguring Urban Governance in Different State Contexts, *Urban Geography* 25, no. 8: 755–770.

Emerson, Kirk, Tina Nabatchi, and Stephen Baloch. 2012. An Integrated Framework for Collaborative Governance. *Journal of Public Administration Research and Theory* 22, no. 1 (January): 1–30. https://doi.org/10.1093/jopart/mur011.

Enas, Alhassan, R. Sandra Schillo, Margaret A. Leamay, and Fred Pries. 2019. Research Outputs as Vehicles of Knowledge Exchange in a Quintuple Helix Context: The Case of Biofuels Research Outputs. *Journal of the Knowledge Economy* 10, no. 3 (September): 958–973.

Etzkowitz, Henry. 1993. Technology Transfer: The Second Academic Revolution. *Technology Access Report* 6, 7–9.

Etzkowitz, Henry. 2003. Research Groups as "Quasi-Firms": The Invention of the Entrepreneurial University. *Research Policy* 32, no. 1 (January): 109–121. https://doi.org/10.1016/S0048-7333(02)00009-4.

Etzkowitz, Henry, and Loet Leydesdorff. 1995. The Triple Helix—University-Industry-Government Relations: A Laboratory for Knowledge Based Economic Development. *EASST Review* 14, no. 1 (January): 14–19.

Etzkowitz, Henry and Loet Leydesdorff. 2000. The Dynamics of Innovation: From National Systems and "Mode 2" to a Triple Helix of University–Industry–Government Relations. *Research Policy* 29: 109–123. https://doi:10.1016/S0048-7333(99)00055-4.

Eurocities. 2011. *Social Economy in Cities: Bologna*. Eurocities Network of Local Authority on Active Inclusion (Eurocities-NLAO). http://nws.eurocities.eu/MediaShell/media/LAO%20Bologna_Social_Economy.pdf.

Europapress. 2019. PP y Cs derogarán en 2020 la ordenanza de Cooperación Público-Social, muro contra la arbitrariedad, alerta Más Madrid. *Europapress*. December 27, 2019. https://www.europapress.es/madrid/noticia-pp-cs-derogaran-2020-ordenanza-cooperacion-publico-social-muro-contra-arbitrariedad-alerta-mas-madrid-20191227154945.html.

European Investment Bank (EIB). 2018. 2019 European Fund for Strategic Investments Report. *European Investment Bank*. https://www.eib.org/attachments/strategies/efsi_2019_report_ep_council_en.pdf.

Farrell, Joseph, and Matthew Rabin. 1996. Cheap Talk. *The Journal of Economic Perspectives* 10, no. 3 (Summer): 103–118. https://doi.org/10.1257/jep.10.3.103.

Federal Constitutional Court, Order of the First Senate of Mar. 24, 2021, 1 BvR 2656/18. Accessed March 21, 2022. http://www.bverfg.de/e/rs20210324_1bvr265618en.html.

Federici, S. 2018. Re-enchanting The World: Feminism and the Politics of the Commons. PM Press.

Fennell, Lee. 2015. Agglomerama. *Brigham Young University Law Review* 2014, no. 6: 1373–1414.

Fennell, Lee. 2016. Fee Simple Obsolete. *New York University Law Review* 91, no. 6 (December): 1457–1516. https://chicagounbound.uchicago.edu/journal_articles.

Fernandes, Edesio. 2007. Constructing the Right to the City in Brazil. *Social & Legal Studies* 16, no. 2 (June): 201–219. https://doi.org/10.1177/0964663907076529.

Fernandes, Edesio. 2011. Implementing the Urban Reform Agenda in Brazil: Possibilities, Challenges, and Lessons. *Urban Forum* 22, no. 3 (June): 229–314.

Ferrero, Giovanni, and Alice Zanasi. 2020. Co-City Torino. *Urban Maestro* (website). https://urbanmaestro.org/wp-content/uploads/2020/09/urban-maestro_co-city-torino_g-ferrero-a-zanasi.pdf.

Finck, Michele, and Sofia Ranchordás. 2016. Sharing and the City. *Vand. J. Transnat'l L* 49: 1299.

Finholt, Thomas. 2002. Collaboratories. *Annual Review of Information Science and Technology* 36, no. 1: 73–107.

Fink, Larry. 2019. Profit and Purpose. *BlackRock*. https://www.blackrock.com/americas-offshore/2019-larry-fink-ceo-letter.

Fink, Larry. 2020. A Fundamental Reshaping of Finance. Letter to CEOs. *BlackRock.* https://www.blackrock.com/uk/individual/larry-fink-ceo-letter.

Florida, Richard L. 2002. *The Rise of the Creative Class and How It's Transforming Work, Leisure, Community and Everyday Life.* New York: Basic Books.

Florida, Richard. (2012) 2017. *The Rise of the Creative Class: Revisited.* New York: Basic Books.

Florida, Richard. 2017a. The "Big Liberal City" Isn't Big Enough. *Bloomberg CityLab* (blog). March 30, 2017. http://www.citylab.com/life/2017/03/the-big-liberal-city-isnt-big-enough/521094.

Florida, Richard. 2017b. *The New Urban Crisis: How Our Cities Are Increasing Inequality, Deepening Segregation, and Failing the Middle Class—And What We Can Do About It.* New York: Basic Books.

Florida, Richard. 2019. The Changing Geography of America's Creative Class. *Bloomberg CityLab* (blog). August 27, 2019. https://www.bloomberg.com/news/articles/2019-08-27/the-changing-geography-of-america-s-creative-class.

Foley, Rider W., Olivier Sylvain, and Sheila Foster. 2022. Innovation and Equality: An Approach to Constructing a Community Governed Network Commons. *Journal of Responsible Innovation.* https://doi:10.1080/23299460.2022.2043681.

Fondazione Innovazione Urbana. n.d.a. Attivazione di Percorsi di Partecipazione e Coproduzione. Last modified August 13, 2020. http://www.fondazioneinnovazioneurbana.it/immaginazionecivica.

Fondazione innovazione Urbana. n.d.b. *Collaborare è Bologna, Fondazione Innovazione Urbana* (website). Accessed March 20, 2022. https://www.fondazioneinnovazioneurbana.it/68-urbancenter/collaborare-bologna.

Foster, Sheila Rose. 2006. The City as an Ecological Space: Social Capital and Urban Land Use. *Notre Dame Law Review* 82, no. 2 (December): 101–152. https://ssrn.com/abstract=899617.

Foster, Sheila Rose. 2011. Collective Action and the Urban Commons. *Notre Dame Law Review* 87, no. 1 (September): 57–134.

Foster, Sheila Rose, and Daniel Bonilla. 2011. Introduction to Symposium. *Fordham Law Review* 80, no. 3: 1003–1015.

Foster, Sheila Rose, and Christian Iaione. 2016. The City as a Commons. *Yale Law and Policy Review* 34, no. 2 (July): 281–349.

Foster, Sheila Rose, and Christian Iaione. 2019. Ostrom in the City: Design Principles and Practices for the Urban Commons. In *Routledge Handbook of The Study of the Commons*, edited by Blake Hudson, Jonathan Rosenbloom, and Dan Cole. Oxfordshire, UK: Routledge, Taylor & Francis.

Franceys, Richard, and Almud Weitz. 2003. Public-Private Community Partnerships in Infrastructure for the Poor. *Journal of International Development* 15, no. 8 (November): 1083–1098.

Franz, Yvonne. 2015. Designing Social Living Labs in Urban Research. *Info* 17, no. 4 (June): 53–66. https://doi.org/10.1108/info-01-2015-0008.

Fransen Lieve, Gino del Bufalo Gino, and Edoardo Reviglio. 2018. Boosting Investment in Social Infrastructure in Europe, Report of the HLTF Force on Investing in Social Infrastructure in Europe chaired by Romano Prodi and Christian Sautter, Discussion Paper, 074, January. https://ec.europa.eu/info/sites/default/files/economy-finance/dp074_en.pdf.

Freeman, Jody. 1997. Collaborative Governance in the Administrative State. *UCLA Law Review* 45, no. 1: 1–98.

Freeman, Jody. 2000. The public role in private governance. *New York University Law Review* 75, no. 3: 543–675.

Frey, William H. 2011. Young Adults Choose "Cool Cities" During Recession. *Brookings Institution Blog*. October 28, 2011. https://www.brookings.edu/blog/up-front/2011/10/28/young-adults-choose-cool-cities-during-recession/.

Frey, William H. 2012. Demographic Reversal: Cities Thrive, Suburbs Sputter. *Brookings Institution Blog*. June 19, 2012. https://www.brookings.edu/opinions/demographic-reversal-cities-thrive-suburbs-sputter/.

Friendly, Abigail. 2013. The Right to the City: Theory and Practice in Brazil. *Planning Theory & Practice*, 14, no. 2 (April): 158–179.

Frischmann, Brett. 2012. *Infrastructure: The Social Value of Shared Resources*. Oxford: Oxford University Press. https://doi.org/10.1093/acprof:oso/9780199895656.001.0001.

Frischmann, Brett M., Michael J. Madison, and Katherine J. Strandburg. 2014. Governing Knowledge Commons. In *Governing Knowledge Commons*. Oxford: Oxford University Press.

Frug, Gerald E. 2001. *City Making: Building Communities Without Building Walls*. Princeton, NJ: Princeton University Press.

Fung, Archon. 2001. Accountable Autonomy: Toward Empowered Deliberation in Chicago Schools and Policing. *Politics and Society* 29, no. 1 (March): 73–103. https://doi.org/10.2139/ssrn.253840.

Fung, Archon. 2004. *Empowered Participation: Reinventing Urban Democracy*. Princeton, NJ: Princeton University Press.

Fung, Dilly. 2016. Engaging Students with Research Through a Connected Curriculum: An Innovative Institutional Approach. *Council on Undergraduate Research Quarterly* 37, no. 2 (Winter): 30–35.

Fung, Dilly. 2017. *A Connected Curriculum for Higher Education*. London: UCL Press.

Galanti, Maria Tullia. 2014. From Leaders to Leadership. Urban Planning, Strategic Exchanges and Policy Leadership in Turin (1993–2011). *Rivista italiana di scienza politica*, 2 (August): 147–174.

Galizzi, Paolo, and Ernest K. Abotsi. 2011. Traditional Institutions and Governance in Modern African Democracies. In *The Future of African Customary Law*, edited by Jeanmarie Fenrich, Paolo Galizzi, and Tracy Higgins. Cambridge: Cambridge University Press.

García-López, Gustavo A., and Camille Antinori. 2017. Between Grassroots Collective Action and State Mandates: The Hybridity of Multi-Level Forest Associations in Mexico. *Conservation and Society* 16, no. 2: 193–204. http://www.jstor.org/stable/26393329.

Gibbons, Michael, ed. 1994. *The New Production of Knowledge: The Dynamics of Science and Research in Contemporary Societies*. London: Thousand Oaks, Calif: SAGE Publications.

Gibson, Clark, Elinor Ostrom, and T.K. Ahn. 2000. The Concept of Scale and the Human Dimension of Global Change: A Survey. *Ecological Economy* 32: 217–39.

Ginocchini, Giovanni, Veronica Conte, Ilaria Daolio, and Irene Sensi. 2013. Passagio A Nordest. Bologna, Italy: Urban Center Bologna and the City of Bologna. https://www.fondazioneinnovazioneurbana.it/images/areeannessesud/report_intermedio_pilastro_public.pdf.

Ginocchini, Giovanni, and Elena Vai, eds. 2021. *Visioni e Azioni dell'istituzione dedicata alle trasformazioni di Bologna. Fondazione Innovazione Urbana*. https://issuu.com/urbancenterbologna/docs/pubblicazione_fiu_pagine_singole.

Gioia, Patrizia. 2004. *Centocelle: Roma S.D.O. le Indagini Archeologiche, Volume I*. Soveria Mannelli, Italy: Rubbettino.

Glaeser, Edward L. 1998. Are Cities Dying? *Journal of Economic Perspectives* 12, no. 2 (Spring): 139–160.

Glaeser, Edward L. 1999. Learning in Cities. *Journal of Urban Economics* 46, no. 2 (September): 254–277

Glaeser, Edward L., Jed Kolko, and Albert Saiz. 2001. The Consumer City. *Journal of Economic Geography* 1, no. 1 (January): 27–50.

Goldsmith, Stephen, and Susan Crawford. 2014. *The Responsive City: Engaging Communities Through Data-Smart Governance*, Hoboken, NJ: Jossey-Bass.

Goodman, Ellen P., and Julia Powles. 2019. Urbanism Under Google: Lessons from Sidewalk Toronto. *Fordham Law Review* 88, no. 2: 457–498.

Grafton, Quentin R. 2000. Governance Of the Commons: A New Role for The State? *Land Economics* 76, no.4 (November): 504–517. https://doi.org/10.2307/3146949.

Great Cities Institute. 2020. *Participatory Budgeting.* University of Illinois at Chicago (website). Accessed January 12, 2020. https://greatcities.uic.edu/uic-neighborhoods-initiative/participatory-budgeting/.

Grignani, Anna, Michela Gozzellino, Alessandro Sciullo, and Dario Padovan. 2021. Community Cooperative: A New Legal Form for Enhancing Social Capital for the Development of Renewable Energy Communities in Italy. *Energies* 14, no. 21: 7029. https://doi.org/10.3390/en14217029.

Grimes, Howard D. 2016. Creating a "Collaboratory" Environment to Transcend Traditional Research Barriers: Insights from the United States. *Energy Research & Social Science* 19, (September): 37–38.

Gross, Jill Simone. 2005. Business Improvement Districts in New York City's Low-Income and High-Income Neighborhoods. Economic Development Quarterly 19, no. 2 (May):174–189. https://doi.org/10.1177/0891242404273783.

Gross, Julian, Greg LeRoy, and Madeline Janis-Aparicio. 2005. *Community Benefits Agreements: Making Development Projects Accountable published by Good Jobs First & The California Partnership for Working Families.* Good Jobs First and the California Partnership for Working Families. https://www.forworkingfamilies.org/resources/publications/community-benefits-agreements-making-development-projects-accountable.

Grossi, Paolo. 2017. *Un altro modo di possedere: L'emersione di forme alternative di proprietà alla coscienza giuridica postunitaria.* Milan: Giuffrè.

Guerrini, Federico. 2014. How Seoul Became One of the World's Sharing Capitals. *Forbes,* May 25, 2014. http://www.forbes.com/sites/federicoguerrini/2014/05/25/how-seoul-became-one-of-the-worlds-sharing-capitals/.

Han, Didi K. 2019. Weaving the Common in the Financialized City: A Case of Urban Cohousing Experience in South Korea, In *Neoliberal Urbanism, Contested Cities and Housing in Asia,* edited by Yi-Ling Chen and Hyun Bang Shin, 171–192. New York: Palgrave Macmillan.

Han, Didi K., and Hajime Imamasa. 2015. Overcoming Privatized Housing in South Korea: Looking through the Lens of "Commons" and "the Common." In *Urban Commons: Moving Beyond State and Market,* edited by Mary Dellenbaugh, Markus Kip, Majken Bieniok, Agnes Müller, and Martin Schwegman, 91–100. Basel, Switzerland: Birkhäuser. https://doi.org/10.1515/9783038214953-006.

Hardin, Garrett. 1968. The Tragedy of the Commons. *Science* 162, no. 3859 (December): 1243–1248.

Harding, Rebecca 2009. Fostering University-Industry Links. In *Strengthening Entrepreneurship and Economic Development in East Germany: Lessons from Local Approaches,* edited by the Organisation for Economic Co-operation and Development, OECD. 139–154. http://www.oecd.org/site/cfecpr/42367462.pdf.

Hardt, Michael, and Antonio Negri. 2009. *Commonwealth*. Cambridge, MA: Harvard University Press.

Harman, Ben P, Bruce M Taylor, and Marcus B Lane. 2015. Urban Partnerships and Climate Adaptation: Challenges and Opportunities. *Current Opinion in Environmental Sustainability* 12 (February): 74–79. https://doi.org/10.1016/j.cosust.2014.11.001.

Harrison, Jeffrey S., Robert A. Phillips, and R. Edward Freeman. 2020. On the 2019 Business Roundtable Statement on the Purpose of a Corporation. *Journal of Management* 46, no. 7 (December): 1223–1237

Harvey, David. 1989. From Managerialism to Entrepreneurialism: The Transformation in Urban Governance in Late Capitalism. *Geografiska Annaler: Series B, Human Geography* 71, no. 1: 3–17.

Harvey, David. 2012. *Rebel Cities: From the Right to the City to the Urban Revolution*. London: Verso Press.

Hernández-Torrales, María E., Lyvia Rodríguez Del Valle, Line Algoed, and Karla Torres Sueiro. 2020. Seeding the CLT in Latin America and the Caribbean: Origins, Achievements, and the Proof-of-Concept Example of the Caño Martín Peña Community Land Trust. In *On Common Ground: International Perspectives on The Community Land Trust*, edited by John Emmeus Davis, Line Algoed, and María E. Hernández-Torrales, 189–210. Madison, WI: Terra Nostra Press.

Hess, Charlotte. 2008. *Mapping the New Commons*. Rochester, NY: SSRN. https://dx.doi.org/10.2139/ssrn.1356835.

Hess, Charlotte, and Elinor Ostrom. 2007. *Understanding Knowledge as a Commons*. Boston: MIT Press.

Hollands, Robert G. 2008. Will the Real Smart City Please Stand Up? *City* 12, no. 3 (November): 303–320.

House of Commons, Communities and Local Government Committee. 2015. Community Rights, Sixth Report of Session 2014–15. Ordered by the House of Commons. https://publications.parliament.uk/pa/cm201415/cmselect/cmcomloc/262/262.pdf.

Humboldt-Universität zu Berlin. 2011. *Bildung durch Wissenschaft—Educating Enquiring Minds*. Berlin: Humboldt-Universität. https://www.um.edu.mt/__data/assets/pdf_file/0012/389973/HUB.pdf.

Huron, Amanda. 2015. Working with Strangers in Saturated Space: Reclaiming and Maintaining the Urban Commons. *Antipode* 47, no. 4 (January): 963–979.

Huron, Amanda. 2018. *Carving Out the Commons: Tenant Organizing and Housing Cooperatives in Washington, DC*. Minneapolis: University of Minnesota Press.

Hyra, Derek S. 2012. Conceptualizing the New Urban Renewal: Comparing the Past to the Present. *Urban Affairs Review* 48, no. 4 (July): 498–527. https://doi.org/10.1177/1078087411434905.

Iaione, Christian. 2010. The Tragedy of Urban Roads: Saving Cities from Choking, Calling on Citizens to Combat Climate Change. *Fordham Urban Law Journal* 37, no. 3 (January): 889–951.

Iaione, Christian. 2012. City as a Commons. In *Design and Dynamics of Institutions for Collective Action: A Tribute to Professor Elinor Ostrom–II Thematic Conference of the IASC*. Utrecht, Netherlands: Utrecht University. http://hdl.handle.net/10535/8604.

Iaione, Christian. 2013. La Città Come Bene Comune. *Aedon* 1: 1127–1345. http://www.aedon.mulino.it/archivio/2013/1/iaione.htm.

Iaione, Christian. 2015. Governing the Urban Commons. *Italian Journal of Public Law* 7, no.1 (April): 170–223.

Iaione, Christian. 2016. The Co-City: Sharing, Collaborating, Cooperating, Commoning in the City. *American Journal of Economics and Sociology* 75, no. 2 (March): 415–459. https://ssrn.com/abstract=2708296.

Iaione, Christian. 2018. *The Pacts of Collaboration as Public-People Partnerships*. UIA Cities, Zoom-in I. https://www.uia-initiative.eu/sites/default/files/2018-07/Turin%20-%2001-051%20Co-City%20-%20Christian%20Iaione%20-%20Zoom-in%201-%20July%202018.pdf.

Iaione, Christian. 2019a. Pooling Urban Commons: The Civic eState. *URBACT* (blog), July 16, 2019. https://urbact.eu/urban-commons-civic-estate.

Iaione, Christian. 2019b. Legal Infrastructure and Urban Networks for Just and Democratic Smart Cities. *Italian Journal of Public Law* 11, no. 2: 747–786. Accessed March 12, 2022. http://www.ijpl.eu/archive/2019/issue-28/legal-infrastructure-and-urban-networks-for-just-and-democratic-smart-cities.

Iaione, Christian, Monica Bernardi, and Elena De Nictolis. 2019. *La Casa per Tutti*. Bologna, Italy: il Mulino.

Iaione, Christian, and Paola Cannavò. 2015. The Collaborative and Polycentric Governance of the Urban and Local Commons. In *Urban Pamphleteer no. 5. Global Education for Urban Futures*, edited by Paola A. d'Alençon, Ben Campkin, Rupali Gupte, Mkhabela Solam, Johannes Novy, and Mika Savela, 29–31. London: UCL Urban Laboratory.

Iaione, Christian, and Elena de Nictolis. 2017. Urban Pooling. *Fordham Urban Law Journal* 44, no. 3 (July): 665–701.

Iaione, Christian, and Elena De Nictolis. 2020. The Role of Law in Relation to the New Urban Agenda and the European Urban Agenda: A Multi-Stakeholder Perspective. In *Law and the New Urban Agenda*. New York: Routledge.

Iaione, Christian, and Elena de Nictolis. 2021. The City as a Commons Reloaded: From the Urban Commons to Co-Cities; Empirical Evidence on the Bologna Regulation. In *Cambridge Handbook of Innovations in Commons Scholarship*, edited by Sheila

Foster and Chrystie Swiney. Cambridge: Cambridge University Press. http://dx.doi.org/10.2139/ssrn.3865774.

Iaione, Christian, Elena De Nictolis, and Anna Berti Suman. 2019. The Internet of Humans (IoH): Human Rights and Co-Governance to Achieve Tech Justice in the City. *Law and Ethics of Human Rights* 13, no. 2: 263, 299.

Irazábal, Clara. 2016. Public, Private, People Partnerships (PPPPs): Reflections from Latin American Cases. In *Private Communities and Urban Governance*, edited by Amnon Lehavi, 191–214. Cham: Springer.

Irazábal, Clara. 2018. Counter Land-Grabbing by the Precariat: Housing Movements and Restorative Justice in Brazil. *Urban Science* 2, no. 2 (June): 49. https://doi.org/10.3390/urbansci2020049

Istituto Italiano di statistica (ISTAT). 2017. Commissione parlamentare di inchiesta sulle condizioni di sicurezza e sullo stato di degrado delle città e delle loro periferie. *Istat*, January 24, 2017. https://www.istat.it/it/files//2017/01/A-audizione24_01_17_REV.pdf.

Istituto Italiano di statistica (ISTAT). 2020. Popolazione residente comunale per sesso, anno di nascita e stato civile. *Istat*.

Istituto Italiano di statistica (ISTAT). 2021. Tasso di disoccupazione livello provinciale. *Istat*.

Italiano Legge Constituzionale no.1. 2022. Gazzetta Ufficiale 44 (February 22). https://www.gazzettaufficiale.it/eli/id/2022/02/22/22G00019/sg.

Jacobs, Jane. 1961. *The Death and Life of Great American Cities*. New York: Random House.

Jacobs, Jane. 1969. *The Economy of Cities*. New York: Vintage.

Jessop, Bob. 2016. *The State: Past, Present, Future*. Cambridge, UK: Polity Press.

Johnson, Cat. 2014. Sharing City Seoul: A Model for the World. *Shareable* (blog), June 3, 2014. http://www.shareable.net/blog/sharing-city-seoul-a-model-for-the-wor.

Johnstone-Louis, Mary, Bridget Kustin, Colin Mayer, Judith Stroehle, and Boya Wang. 2020. Business in Times of Crisis. *Oxford Review of Economic Policy* 36, no. 1: S242–S255. https://doi.org/10.1093/oxrep/graa021.

Joss, Simon, Frans Sengers, Daan Schraven, Federico Caprotti, and Youri Dayot. 2019. The Smart City as Global Discourse: Storylines and Critical Junctures across 27 Cities. *Journal of Urban Technology* 26, no. 1 (February): 3–34.

JRC Science Hub Communities. 2020a. *City Science Initiative*. European Commission, JRC Science Hub Communities (website). Accessed September 10, 2021. https://ec.europa.eu/jrc/communities/en/community/city-science-initiative.

JRC Science Hub Communities. 2020b. *Virtual Workshop on Tech and the City*. European Commission, JRC Science Hub Communities (website). Accessed September 10, 2021.

https://ec.europa.eu/jrc/communities/en/community/city-science-initiative/event/virtual-workshop-%C2%A0tech-and-city.

Kaplan, Ivan. 2012. Does the Privatization of Publicly Owned Infrastructure Implicate the Public Trust Doctrine? Illinois Central and the Chicago Parking Meter Concession Agreement. *Northwestern Journal of Law and Social Policy* 7, no. 1 (Winter): 136–169.

Karamarkos, Kostas. 2018. Are Urban Gardens the place for modern community hubs? *UrbAct*, last edited June 24, 2019. https://urbact.eu/are-urban-gardens-place-modern-community-hubs.

Karvonen, Andrew, and Bas van Heur. 2014. Urban Laboratories: Experiments in Reworking Cities. *International Journal of Urban and Regional Research* 38, no. 2 (December): 379–392. https://doi.org/10.1111/1468-2427.12075.

Kim, Hyung M., and Sun S. Han. 2012. City Profile: Seoul. *Cities* 29, no. 2 (April): 142–154. https://doi.org/10.1016/j.cities.2011.02.003.

Kinkead, Eugene. 1990. *Central Park*. New York: W. W. Norton.

Kinney, Jen. 2016. NYC Comptroller Pushes Land Bank as Affordable Housing Tool. *Next City Magazine* (blog), February 19, 2016. https://nextcity.org/daily/entry/new-york-city-land-bank-creation-vacant-properties-affordable-housing.

Kiple. n.d. *Home*. Kiple (website). Accessed September 10, 2021. http://www.kiple.net/.

Kitchin, Rob, Claudio Coletta, Leighton Evans, and Liam Heaphy. 2018. Creating Smart Cities: Introduction. In *Creating Smart Cities*, edited by Claudio Coletta, Leighton Evans, Liam Heaphy, and Rob Kitchin. London: Routledge.

Kitchin, Rob, Claudio Coletta, Leighton Evans, Liam Heaphy, and Darach Mac Donncha. 2017. Smart Cities, Urban Technocrats, Epistemic Communities, Advocacy Coalitions and the "Last Mile" Problem. *IT—Information Technology* 59, no. 6 (November): 275–284.

Kooiman, Jan. 1993. *Modern Governance. New Government-Society Interactions*. London: Sage.

Kooiman, Jan. 2003. *Governing as Governance*. London: Sage.

Kooiman, Jan, and Martijn van Vliet. 2000. Self-Governance as a Mode of Societal Governance. *Public Management an International Journal of Research and Theory* 2, no. 3: 359–378. https://doi.org/10.1080/14719030000000022.

Kooiman, Jan, and Maarten Bavinck. 2005. The Governance Perspective. In *Fish for life. Interactive Governance for Fisheries*, edited by Jan Kooiman, Maarten Bavinck, Svein Tentoft, and Roger Pullin. Amsterdam: Amsterdam University Press.

Kretzmann, J., and J. McKnight. 1993. *Building Communities from the Inside Out: A Path Toward Finding and Mobilizing a Community's Assets*. Evanston, IL: Center for Urban Affairs and Policy Research, Northwestern University.

Laerhoven, Frank van, and Clare Barnes. 2019. Facilitated Self-Governance of the Commons: On the Roles of Civil Society Organizations in the Governance of Shared

Resource Systems. In *Routledge Handbook on the Study of the Commons*, edited by Blake Hudson, Jonathan Rosenbloom, and Dan Cole. New York: Routledge.

Laino, Giovanni. 2012. *Il fuoco nel cuore e il diavolo in corpo*. Milan, Italy: Franco Angeli.

Lang, Richard, Dietmar Roessl, and Daniela Weismeier-Sammer. 2013. Co-operative Governance of Public-Citizen partnerships: Two Diametrical Participation Modes. In *Conceptualizing and Researching Governance in Public and Non-Profit Organizations*, edited by Luca Gnan, Alessandro Hinna, and Fabio Monteduro, 227–246. Bingley, UK: Emerald Group.

Layard, Antonia. 2012. The Localism Act 2011: What is Local and How Do We (Legally) Construct It. *Environmental Law Review* 14, no. 2 (May): 134–144. https://doi.org/10.1350/enlr.2012.14.2.152.

Lazonick, William, and Mariana Mazzucato. 2013. The Risk-Reward Nexus in the Innovation-Inequality Relationship: Who Takes the Risks? Who Gets the Rewards? *Industrial and Corporate Change* 22, no. 4 (July): 1093–1128.

Lee, Caroline W., Michael McQuarrie, and Edward T. Walker. 2015. *Democratizing Inequalities: Dilemmas of the New Public Participation*. New York: New York University Press.

Lee, Ryun Jung, and Galen Newman. 2021. The Relationship Between Vacant Properties and Neighborhood Gentrification. *Land Use Policy* 101 (November): 105185. https://doi.org/10.1016/j.landusepol.2020.105185.

Lefebvre, Henri. (1968) 1996. The Right to City. In *Writings on Cities*. Hoboken, NJ: Wiley-Blackwell. First published as *Le Droit A La Ville*, Paris, France: Anthropos.

Le Feuvre, Meryl Dominic Medway, Gary Warnaby, Kevin Ward, and Anna Goatman. 2016. Understanding Stakeholder Interactions in Urban Partnerships, *Cities* 52: 55–65. https://doi.org/10.1016/j.cities.2015.10.017.

Lehavi, Amnon. 2004. Property Rights and Local Public Goods: Toward a Better Future for Urban Communities. *Urban Lawyer* 36, no. 1 (Winter): 1–99. https://ssrn.com/abstract=694063.

Lehavi, Amnon. 2008. How Property Can Create, Maintain, or Destroy Community. *Theoretical Inquiries in Law* 10(1): 43–76. https://doi.org/10.2202/1565-3404.1208.

Lelo, Keti, Salvatore Monni, and Federico Tomassi. 2018. Roma, tra centro e periferie: come incidono le dinamiche urbanistiche sulle disuguaglianze socio-economiche. *Roma Moderna e Contemporanea* XXV, no. 1–2. (July). https://doi.org/10.17426/6004.

Lelo, Keti, Salvatore Monni, and Federico Tomassi. 2021. *Le sette Rome: La capitale delle disuguaglianze raccontata in 29 mappe*. Roma: Donzelli Editore.Lerner, Josh, and Donata Secondo. 2012. By the People, for the People: Participatory Budgeting from the Bottom Up in North America. *Journal of Public Deliberation* 8, no. 2 (December): Article 2.

Levine, Peter. 2007. Collective Action, Civic Engagement, and the Knowledge Commons. In *Understanding Knowledge as a Commons*, edited by Charlotte Hess and Elinor Ostrom. Cambridge, MA: MIT Press.

Leydesdorff, Loet, and Henry Etzkowitz. 1998. The Triple Helix as a Model for Innovation Studies. *Science and Public Policy* 25, no. 3 (June): 195–203.

Leydesdorff, Loet. 2012. The Triple Helix, Quadruple Helix, . . . , and an N-tuple of Helices: Explanatory Models for Analyzing the Knowledge-Based Economy? *Journal of the Knowledge Economy* 3, no. 1: 25–35.

Lievens, Matthias. 2014. From Government to Governance: A Symbolic Mutation and Its Repercussions for Democracy. *Political Studies* 63, no. 1 (October): 2–17. https://doi.org/10.1111/1467-9248.12171.

Linebaugh, Peter. 2008. *The Magna Carta Manifesto: Liberties and Commons for All*. Berkeley: University of California Press.

Livengood, Chad. 2019. Detroit Strikes Land Deal with Hantz Farms as Part of FCA Plant Project. *Crain's Detroit Business*. April 15, 2019. https://www.crainsdetroit.com/real-estate/detroit-strikes-land-deal-hantz-farms-part-fca-plant-project.

Logan, John R., and Henry Molotch. 1987. *Urban Fortunes: The Political Economy of Place*. Berkeley: University of California Press.

Lucarelli, Alberto. 2011. *Beni comuni. Dalla teoria all'azione politica*. Viareggio, Italy: Dissensi.

Lucarelli, Alberto. 2017. Acqua bene comune (ABC) (Italie). In *Dictionnaire Des Biens Communs*, 1re édition. Quadrige, edited by Marie Cornu, Fabienne Orsi, and Judith Rochfeld. Paris: Presses universitaires de France.

Madden, Kathy. 2000. *Public Parks, Private Partners: How Partnerships Are Revitalizing Urban Parks*. New York: Project for Public Spaces.

Madison, Michael J., Brett M. Frischmann, and Katherine J. Strandburg. 2016. Knowledge Commons. In *Research Handbook on the Economics of Intellectual Property Law (Vol. II—Analytical Methods)*, edited by P. Menell and D. Schwartz. Cheltenham, UK: Edward Elgar.

Madison, Michael J., Brett M. Frischmann, and Katherine J. Strandburg. 2010. Constructing Commons in the Cultural Environment. *Cornell Law Review* 95, no. 4 (May): 657–710. https://scholarship.law.cornell.edu/clr/vol95/iss4/10.

Madison, Michael J., Brett M. Frischmann, and Katherine J. Strandburg. 2014. Governing Knowledge Commons. In *Governing Knowledge Commons*. Oxford: Oxford University Press.

Majamaa, W., S. Junnila, H. Doloi, and E. Niemistö. 2008. End-User Oriented Public-Private Partnerships in Real Estate Industry. *International Journal of Strategic Property Management* 12, no. 1: 1–17. https://doi.org/10.3846/1648-715X.2008.12.1-17.

Mallach, Alan. 2018. *The Empty House Next Door: Understanding and Reducing Vacancy and Hypervacancy in the United States*. Cambridge, MA: Lincoln Institute of Land Policy.

Mansbridge, Jane. 2014. The Role of the State in Governing the Commons. *Environmental Science and Policy* 36 (February): 8–10.

Marana, Patricia, Labaka, Leire and Jose Mari Sarriegi. 2018. A Framework for Public-Private-People Partnerships in the City Resilience-Building Process. *Safety Science* 110, Part C: 39–50. https://doi.org/10.1016/j.ssci.2017.12.011.

Marella, Maria Rosaria. 2017. The Commons as a Legal Concept. *Law and Critique* 28, no. 1 (March): 61–86. https://doi.org/10.1007/s10978-016-9193-0.

Marantz, Nicholas J. 2015. What Do Community Benefits Agreements Deliver? Evidence From Los Angeles. *Journal of the American Planning Association* 81, no. 4: 251–257. http://www.tandfonline.com/doi/full/10.1080/01944363.2015.1092093.

Masella, Nicola. 2018. Urban Policies and Tools to Foster Civic Uses: Naples' Case Study. *Urban Research & Practice* 11, no. 1 (January): 78–84. https://doi.org/10.1080/17535069.2018.1426689.

Mattei, Ugo. 2013. Protecting the Commons: Water, Culture, and Nature. The Commons Movement in the Italian Struggle against Neoliberal Governance. *South Atlantic Quarterly* 112, no. 2: 366–376.

Mattei, Ugo, and Saky Bailey. 2013. Social Movements as Constituent Power: The Italian Struggle for the Commons. *Indiana Journal of Global Legal Studies* 20, no. 2. https://www.repository.law.indiana.edu/ijgls/vol20/iss2/14.

Mattei, Ugo, and Alessandra Quarta. 2014. *L'acqua e il suo diritto*. Rome: Futura Editrice.

Mattei, Ugo, and Alessandra Quarta. 2015. Right to the City or Urban Commoning? Thoughts on the Generative Transformation of Property Law. *Italian Law Journal* 1, no. 2 (March): 303–325. https://repository.uchastings.edu/faculty_scholarship/1287/.

Mazzucato, Mariana. 2015. Building the Entrepreneurial State: A New Framework for Envisioning and Evaluating a Mission-Oriented Public Sector. *The Levy Economics Institute Working Paper Collection*, no. 824. http://www.levyinstitute.org/pubs/wp_824.pdf.

Mazzucato, Mariana. 2018a. Let's Make Private Data into a Public Good. *MIT Technology Review* 121, no. 4 (July). https://www.technologyreview.com/2018/06/27/141776/lets-make-private-data-into-a-public-good/.

Mazzucato, Mariana. 2018b. *Mission-Oriented Research & Innovation in the European Union: A Problem-Solving Approach to Fuel Innovation-Led Growth*. Luxemburg: European Commission Directorate-General for Research and Innovation.

Mazzucato, Mariana, Rainer Kattel, and Josh Ryan-Collins. 2019. Challenge-Driven Innovation Policy: Towards a New Policy Toolkit. *Journal of Industry, Competition and Trade* 20, no. 2 (December): 421–437.

McGinnis, Michael D. 2016. Polycentric Governance in Theory and Practice: Dimensions of Aspiration and Practical Limitations. Working Paper. University of Washington. http://dx.doi.org/10.2139/ssrn.3812455.

McLaren, Duncan, and Julian Agyeman. 2015. *Sharing Cities: A Case for Truly Smart and Sustainable Cities*. Cambridge, MA: MIT Press.

Medoff, Peter, and Holly Sklar. 1994. *Streets of Hope: The Fall and Rise of an Urban Neighborhood*. Troy, NY: South End Press.

Meloni, Claudia, Cappellaro, Francesca, Chiarini, Roberta and Claudia Snels. 2019. Energy Sustainability and Social Empowerment: The Case of Centocelle Smart Community Co-creation. *International Journal of Sustainable Energy Planning and Management* 24. https://doi.org/10.5278/ijsepm.3339.

Mendell, M., et al. 2010. Improving Social Inclusion at the Local Level through the Social Economy: Report for Korea, OECD Local Economic and Employment Development (LEED) Papers, No. 2010/15, OECD Publishing, Paris. https://doi.org/10.1787/5kg0nvg4bl38-en.

Mendell, M., and Béatrice Alain. 2015. Enabling the Social and Solidarity Economy through the Co-Construction of Public Policy. In *Social and Solidarity Economy Beyond the Fringe*, edited by Peter Utting, 166–182. London: Zed Books.

Mercat Social. 2020. *El Balanç Comunitari*. Mercat Social (website). Accessed September 10, 2021. https://mercatsocial.xes.cat/ca/eines/el-balanc-comunitari/.

Micciarelli, Giuseppe. 2021. *Commoning: Beni comuni, nuove istituzioni—Materiali per una teoria dell'autorganizzazione*. 2nd ed. Naples: Editoriale Scientifica.

Micciarelli, Giuseppe. 2022. Hacking the Legal. The Commons between the Governance Paradigm and Inspirations from the 'Living History' of Collective Land Use. In *Post-Growth Planning: Towards an Urbanisation beyond the Market Economy*, edited by Federico Savini, Antònio Ferreira, and Kim C. von Schönfeld. London: Routledge.

Micciarelli, Giuseppe, and Margherita D'Andrea. 2020. Music, Art, the Power and the Capital: A Theoretical Proposal for an Income of Creativity and Care. In *Commons between Dream and Reality*, edited by Maria Francesca De Tullio, 135–163. Košice, Slovakia: Creative Industry Košice.

Miller, Stephen. 2013. Community Land Trusts: Why Now Is the Time to Integrate This Housing Activists' Tool into Local Government Affordable Housing Policies. In *Zoning & Planning Law Report 1* 36, no. 9 (October).

Mills, Stuart. 2020. Who Owns the Future? Data Trusts, Data Commons, and the Future of Data Ownership. *Future Economies Research and Policy Paper*, no. 7 (January). https://www.mmu.ac.uk/media/mmuacuk/content/documents/business-school/future-economies/Mills-2020.pdf.

Mise Gov. n.d. *Programma di supporto alle tecnologie emergenti 5G*. MiseGov (website). https://www.mise.gov.it/index.php/it/comunicazioni/servizi-alle-imprese/tecnologia-5g/tecnologie-emergenti-5g#asse-I.

Mitchell, Thomas W. 2005. Destabilizing the Normalization of Rural Black Land Loss: A Critical Role for Legal Empiricism. *Wisconsin Law Review* 2005, no. 2 (March): 557–615. https://ssrn.com/abstract=896759.

Montalban, Matthieu, Vincent Frigant, and Bernard Jullien. 2019. Platform Economy as a New Form of Capitalism: A Régulationist Research Programme. *Cambridge Journal of Economics* 43, no. 4 (July): 805–824.

Morckel, Victoria C. 2013. Empty Neighborhoods: Using Constructs to Predict the Probability of Housing Abandonment. *Housing Policy Debate* 23, no. 3 (May): 469–496.

Moretti, Enrico. 2012. *The New Geography of Jobs*. Boston: Houghton Mifflin Harcourt.

Morgera, Elisa. 2016. The Need for an International Legal Concept of Fair and Equitable Benefit Sharing. *European Journal of International Law* 27, no. 2: 353–383. https://doi.org/10.1093/ejil/chw014.

Morlino, Leonardo. 2011. *Changes for Democracy: Actors, Structures, Processes*. Oxford: Oxford University Press.

Mori, Pier Angelo. 2014, Community and Cooperation: The Evolution of Cooperatives Towards New Models of Citizens' Democratic Participation in Public Services Provision. *Euricse Working Papers* 1463. https://ideas.repec.org/p/trn/utwpeu/1463.html.

Mumford, Lewis. 1938. *The Culture of Cities*. New York: Harcourt.

Mundoli, S., B. Manjunath, and H. Nagendra. 2015. Effects of Urbanisation on the Use of Lakes as Commons in the Peri-Urban Interface of Bengaluru, India. *International Journal of Urban Sustainable Development* 7, no. 1: 89–108. https://doi.org/10.1080/19463138.2014.982124.

Murphy, Ailbhe, Johan P. Enqvist, and Maria Tengö. 2019. Place-Making to Transform Urban Social-Ecological Systems: Insights from the Stewardship of Urban Lakes in Bangalore, India. *Sustainability Science* 14 (February): 607–623.

Murray, J. Haskell. 2012. Choose Your Own Master: Social Enterprise, Certifications, and Benefit Corporation Statutes. *American University Business Law Review* 2, no. 1: 1–54. https://digitalcommons.wcl.american.edu/aublr/vol2/iss1/1/.

Murray, Michael. 2010. Private Management of Public Spaces: Nonprofit Organizations and Urban Parks. *Harvard Environmental Law Review* 34: 179–255. https://ssrn.com/abstract=1338583.

Nagendra, Harini, and Elinor Ostrom. 2014. Applying the Social-Ecological System Framework to the Diagnosis of Urban Lake Commons in Bangalore, India. *Ecology and Society* 19, no. 2: 67. http://dx.doi.org/10.5751/ES-06582-190267.

Nagy, Boldizsar, 2009. Speaking Without a Voice. In *Future Generations and International Law*, edited by Emmanuel Agius and Salvino Busuttil, 51–63. London: Earthscan Publications Ltd.

National Public Radio (NPR). 2020. Moms 4 Housing Celebrate Win in Battle Over Vacant House. *Morning Edition* (radio broadcast). January 22, 2020. https://www.npr.org/2020/01/22/798392207/moms-4-housing-celebrate-win-in-battle-over-vacant-house.

National Public Radio (NPR), Empty Houses Reclaimed. 2021. *Planet Money* (radio broadcast). March 1, 2021. https://www.npr.org/transcripts/971873769.

Nesti, Giorgia. 2018. Co-production for Innovation: The Urban Living Lab Experience. *Policy and Society* 37, no. 3: 310–325. https://doi.org/10.1080/14494035.2017.1374692.

Nevejan Caroline. 2020. City Science. In *Values for Survival Cahier* 1, no. 4: 126–131, Rotterdam: Het Nieuwe Instituut.

Nevejan, Caroline, and Frances Brazier. 2015. Design for the Value of Presence. In *Handbook of Ethics, Values, and Technological Design: Sources, Theory, Values and Application Domains*, edited by Jeroen van den Hoven, Pieter E. Vermaas, and Ibo van de Poel. Dordrecht, Netherlands: Springer.

Nicholls, Alex. 2009. "We do good things, don't we?": "Blended Value Accounting" in social Entrepreneurship. *Accounting, Organizations and Society* 34, no. 6–7: 755–769. https://doi.org/10.1016/j.aos.2009.04.008.

Nielsen, Greg. 2015. Mediated Exclusions from the Urban Commons: Journalism and Poverty. In *Urban Commons: Rethinking the City*, edited by Christian Borch and Martin Kornberger. New York: Routledge.

Oakerson, Ronald J., and Jeremy D. W. Clifton. 2017. The Neighborhood as Commons: Reframing Neighborhood Decline. *Fordham Urban Law Journal* 44, no. 2 (May): 411–450.

Olajide, Oluwafemi, and Taibat Lawanson. 2021. Urban Paradox and the Rise of the Neoliberal City: Case Study of Lagos, Nigeria. *Urban Studies* (June). http://doi:10.1177/00420980211014461.

Oliver, Miquel, Johan Zuidweg, and Michail Batikas. 2010. Wireless Commons against the Digital Divide. *International Symposium on Technology and Society*. http://doi:10.1109/ISTAS.2010.5514608.

Olivotto, Veronica. The beginning of the first Co-City: CO-Bologna. TransitSocialInnovation theory (website). Accessed January 2022. http://www.transitsocialinnovation.eu/sii/ctp/ctp4-the-beginning-of-the-first-co-city-co-bologna.

Osayimwese, Itohan, and David Rifkind. 2014. Building Modern Africa: Theme Editors' Introduction, *Journal of Architectural Education* 68, no. 2:156–158. https://doi.org/10.1080/10464883.2014.943631

Ostrom, Elinor. 1990. *Governing the Commons: The Evolution of Institutions for Collective Action*. London: Cambridge University Press.

Ostrom, Elinor. 1998. A Behavioral Approach to the Rational Choice Theory of Collective Action. *American Political Science Review* 92, no. 1 (March): 1–22. https://doi.org/10.2307/2585925.

Ostrom, Elinor. 2000. Private and Common Property Rights. In *Encyclopedia of Law and Economics*, edited by Boudewijn Bouckaert and Gerrit De Geest, 332–379. Cheltenham: Edward Elgar.

Ostrom, Elinor. 2005a. *Understanding Institutional Diversity*. Princeton, NJ: Princeton University Press.

Ostrom, Elinor. 2005b. Unlocking Public Entrepreneurship and Public Economies. Working Paper DP2005/01, World Institute for Development Economic Research (UNU-WIDER). https://www.wider.unu.edu/publication/unlocking-public-entrepreneurship-and-public-economies.

Ostrom, Elinor. 2007. A Diagnostic Approach for Going Beyond Panaceas. *Proceedings of the National Academy of Sciences* 104, no. 39 (July): 15181–15187. https://doi.org/10.1073/pnas.0702288104.

Ostrom, Elinor. 2009a. A General Framework for Analyzing Sustainability of Social-Ecological Systems. *Science* 325, no. 5939 (July): 419–422. https://doi.org/10.1126/science.1172133.

Ostrom, Elinor. 2010a. Beyond Markets and States: Polycentric Governance of Complex Economic Systems. *American Economic Review* 100, no. 3 (June): 641–72.

Ostrom, Elinor. 2010b. Polycentric Systems for Coping with Collective Action and Global Environmental Change. *Global Environmental Change* 20, no. 5 (October): 550–557. https://doi.org/10.1016/j.gloenvcha.2010.07.004.

Ostrom, Elinor. 2011. Background on the Institutional Analysis and Development Framework. *The Policy Studies Journal* 39, no. 1 (February): 7–27. https://doi.org/10.1111/j.1541-0072.2010.00394.x.

Ostrom, Elinor. 2014. A Frequently Overlooked Precondition of Democracy: Citizens Knowledgeable about and Engaged in Collective Action. In *Elinor Ostrom and the Bloomington School of Political Economy: Polycentricity in Public Administration and Political Science*, edited by Daniel H. Cole and Michael G. McGinnis, 337–352. Lanham, MD: Lexington Books.

Ostrom, Elinor, Roger B. Parks, and Gordon P. Whitaker. 1973. Do We Really Want to Consolidate Urban Police Forces? A Reappraisal of Some Old Assertions. *Public Administration Review* 33, no. 5 (September-October): 423–32.

Ostrom, Vincent, Charles M. Tiebout, and Robert Warren. 1961. The Organization of Government in Metropolitan Areas: A Theoretical Inquiry. *American Political Science Review* 55, no. 4 (December): 831–842. https://EconPapers.repec.org/RePEc:cup:apsrev:v:55:y:1961:i:04:p:831-842_12.

Ouellette, Lisa Larrimore. 2015. Patent Experimentalism. *Virginia Law Review* 101, no. 1 (March): 65–128. http://dx.doi.org/10.2139/ssrn.2294774.

Pact of Amsterdam. Urban Agenda for the EU. 2016. Agreed at the Informal Meeting of EU Ministers Responsible for Urban Matters on 30 May 2016 in Amsterdam, The Netherlands. https://ec.europa.eu/regional_policy/sources/policy/themes/urban-development/agenda/pact-of-amsterdam.pdf.

Pape, Madeleine, and Chaeyoon Lim. 2019. Beyond the "Usual Suspects?" Reimagining Democracy with Participatory Budgeting in Chicago. *Sociological Forum* 34, no.4 (September): 861–882.

Participatory Budgeting Project. n.d. *How Participatory Budgeting Works*. Participatory Budgeting Project (website). Accessed September 10, 2021. https://www.participatorybudgeting.org/how-pb-works/.

Patti, Daniela, and Levente Polyák. 2017. *Funding the Cooperative City: Community Finance and the Economy of Civic Spaces*. Vienna, Austria: Cooperative City Books.

PB Chicago. 2020. *What Is Participatory Budgeting?* PB Chicago (website). Accessed September 10, 2021. http://www.pbchicago.org/about.html.

Peñalver, Eduardo Moises, and Sonia K. Katyal. 2010. *Property Outlaws: How Squatters, Pirates, And Protestors Improve the Law of Ownership*. New Haven, CT: Yale University Press.

Perry, Mark J. 2018. Many U.S. Metro Areas Have Greater GDP than Entire Developed Nations. *Foundation for Economic Education* (blog). October 2, 2018. https://fee.org/articles/many-us-metro-areas-have-greater-gdp-than-entire-developed-nations/.

Phillips, Susan. 2020. Homeless mothers squat Federal Housing Sites as Encampment Deadline Looms. *WHYY Opinion* (blog). September 8, 2020. https://whyy.org/articles/homeless-mothers-squat-federal-housing-sites-as-encampment-deadline-looms/.

Pierk, Simone, and Maria Tysiachniouk. 2016. Structures of Mobilization and Resistance: Confronting the Oil and Gas Industries in Russia. *The Extractive Industries and Society* 3, no. 4: 997–1009. https://doi.org/10.1016/j.exis.2016.07.004.

Platform on Sustainable Finance. 2022. Final Report on Social Taxonomy. European Commission (website). Accessed March 13, 2022. https://ec.europa.eu/info/sites/default/files/business_economy_euro/banking_and_finance/documents/280222-sustainable-finance-platform-finance-report-social-taxonomy.pdf.

Polanyi, Karl. 1944. *The Great Transformation: The Political and Economic Origins of Our Time*. Boston: Beacon Press.

Poteete, Amy, and Ostrom Elinor 2004a. In Pursuit of Comparable Concepts and Data about Collective Action. *Agricultural Systems* 82, no. 3 (December): 215–232. https://EconPapers.repec.org/RePEc:eee:agisys:v:82:y:2004:i:3:p:215-232.

Poteete, Amy R., and Elinor Ostrom. 2004b. Heterogeneity, Group Size, and Collective Action: The Role of Institutions in Forest Management. *Development and Change* 35, no. 3 (July): 435–461. https://doi.org/10.1111/j.1467-7660.2004.00360.x.

Poteete, Amy R., Marco A. Janssen, and Elinor Ostrom. 2010. *Working Together: Collective Action, the Commons and Multiple Methods in Practice*. Princeton, NJ: Princeton University Press.

Pritchett, Wendell, and Shitong Qiao. 2018. Exclusionary Megacities. *Southern California Law Review* 9, no. 3 (March): 467–522.

Prno, Jason, and Scott Slocombe. 2012. Exploring the Origins of "Social License to Operate" in the Mining Sector: Perspectives from Governance and Sustainability Theories. *Resources Policy* 37, no. 3: 346–357. https://doi.org/10.1016/j.resourpol.2012.04.002.

Purcell, Mark. 2002. Excavating Lefebvre: The Right to the City and its Urban Politics of the Inhabitant. *GeoJournal* 58 (October): 99–108.

Purcell, Mark. 2013. Possible Worlds: Henri Lefebvre and the Right to the City. *Journal of Urban Affairs*, 36, no. 1 (November): 141–154.

Putnam, Robert, Robert Leonardi, and Raffaella Nanetti. 1993. *Making Democracy Work: Civic Traditions in Modern Italy*. Princeton, NJ: Princeton University Press.

Rae, Douglas. 2003. *City: Urbanism and Its End*. New Haven, CT: Yale University Press. https://www.jstor.org/stable/j.ctt1np937.

Raffol, Matthew. 2012. Community Benefit Agreements in the Political Economy of Urban Development. *Advocates Forum*. Chicago: University of Chicago. Ranchordás, Sofia. 2015. Innovation Experimentalism in the Age of the Sharing Economy. *Lewis & Clark Law Review* 19: 871.

Ratti, Carlo, and Anthony Townsend. 2011. The Social Nexus. *Scientific American* 305, no. 3 (September): 42–49.

Rausch, Stephen, and Cynthia Negrey. 2006. Does the Creative Engine Run? A Consideration of the Effect of Creative Class on Economic Strength and Growth. *Journal of Urban Affairs* 28, no. 5 (November): 473–489.

Raven, Rob, Frans Sengers, Philipp Spaeth, Linjun Xie, Ali Cheshmehzangi, and Martin de Jong. 2017. Urban Experimentation and Institutional Arrangements. *European Planning Studies* 27, no. 2 (October): 258–281. https://doi.org/10.1080/09654313.2017.1393047.

Rawls, John. 1971. *A Theory of Justice*. Boston: Harvard University Press.

Regione Emilia-Romagna. 2010. *Norme per la Definizione, riordino e promozione delle Procedure di Consultazione e Partecipazione alla Elaborazione delle Politiche Regionali e Locali*. Regional Law, approved on February 9, 2010. https://demetra.regione.emilia-romagna

.it/al/articolo?urn=urn:nir:regione.emilia.romagna:legge:2010-02-09;3&urn_tl=dl&urn_t=text/xml&urn_a=y&urn_d=v&urn_dv=n.

Ribera-Fumaz, Ramon. 2019. Moving from Smart Citizens to Technological Sovereignty? In *The Right to the Smart City*, edited by Paolo Cardullo, Cesare Di Feliciantonio, and Rob Kitchin, 177–191. Bingley, UK: Emerald.

Richardson, Harry W., and Chang Woon Nam. 2014. *Shrinking Cities: A Global Perspective*. London: Routledge.

Rifkin, Jeremy. 2014. *The Zero Marginal Cost Society: The Internet of Things, the Collaborative Commons, and the Eclipse of Capitalism*. London: Palgrave Macmillan.

Rinne, April. 2020. Coronavirus: The End of the Sharing Economy, or a New Beginning? *Medium* (website). April 5, 2020. Accessed March 13, 2022. https://medium.com/swlh/coronavirus-the-end-of-the-sharing-economy-or-a-new-beginning-a142acbb7130.

Rodriguez, Daniel B., and David Schleicher. 2012. The Location Market. *George Mason Law Review* 19, no. 3 (April): 637–664.

Rodrik, Dani. 2004. Industrial Policy for the Twenty-First Century, *CEPR Discussion Paper*, (4767). https://tinyurl.com/y9yjzpdq.

Rogers, David. 1990. Community Control and Decentralization. In *Urban Politics New York Style*, edited by Bellush Jewel and Dick Netzer, 143–147. New York: Routledge.

Rogge, Nicole, and Insa Theesfeld. 2018. Categorizing Urban Commons: Community Gardens in the Rhine-Ruhr Agglomeration, Germany. *International Journal of the Commons* 12, no. 2 (October): 251–274. http://doi.org/10.18352/ijc.854.

Rose, Carol. 1986. The Comedy of the Commons: Commerce, Custom and Inherently Public Property. *University of Chicago Law Review* 53, no. 3 (Summer): 711–781.

Rosenzweig, Roy, and Elizabeth Blackmar. 1992. *The Park and the People: A History of Central Park*. Ithaca, NY: Cornell University Press.

Rossi, Ugo, and Alberto Vanolo. 2015. Urban Neoliberalism. *International Encyclopedia of the Social & Behavioral Sciences* 2: 846–853.

Rossini, Luisa, and Iolanda Bianchi. 2019. Negotiating (Re)appropriation Practices Amid Crisis and Austerity. *International Planning Studies* 25, no. 1 (December): 100–121. https://doi.org/10.1080/13563475.2019.1701424.

Rothstein, Richard. 2017. *The Color of Law: A Forgotten History of How Our Government Segregated America*. New York: Liveright.

Sabel, Charles. 2004. *Beyond Principal-Agent Governance: Experimentalist Organizations, Learning and Accountability*, edited by Jeffrey Neil Gordon and Mark J. Roe. Cambridge: Cambridge University Press, 310–327.

Safransky, Sara. 2017. Rethinking Land Struggle in the Postindustrial City. *Antipode* 49, no. 4, (March): 1079–1100. https://doi.org/10.1111/anti.12225.

Salkin, Patricia E., and Amy Lavine. 2008. Negotiating for Social Justice and the Promise of Community Benefits Agreements: Case Studies of Current and Developing Agreements. *Journal of Affordable Housing & Community Development Law* no. 1/2 (Fall/Winter): 113–144.

Samuels, Alec. 2017. Assets of Community Value. *Journal of Planning & Environmental Law* 5, (February): 482–486.

Sarker, Ashutosh. 2013. The Role of State-Reinforced Self-Governance in Averting the Tragedy of the Irrigation Commons in Japan: State-Reinforced Self-Governance. *Public Administration* 91, no. 3 (February): 727–743. https://doi.org/10.1111/padm.12011.

Sassen, Saskia. 2014. *Expulsions: Brutality and Complexity in the Global Economy*. Cambridge, MA: Harvard University Press.

Sassen, Saskia. 2015. Who Owns Our Cities—And Why This Urban Takeover Should Concern Us All. *Guardian* (US edition), November 24, 2015. https://www.theguardian.com/cities/2015/nov/24/who-owns-our-cities-and-why-this-urban-takeover-should-concern-us-all?CMP=share_btn_tw\.

Sax, Joseph L. 1970. The Public Trust Doctrine in Natural Resources Law: Effective Judicial Intervention. *Michigan Law Review* 68, no. 3: 471–566. https://repository.law.umich.edu/mlr/vol68/iss3/3.

Schedule of Clauses Regulating the Concession for Private Use of Public Domain of the Properties Located in the Can Batlló-Magòria Area to the Can Batlló Self-Managed Community and Neighbourhood Space Association, Dossier n. DP-2018–27309. Directorate of Heritage Legal Services (website). Accessed March 13, 2022. http://hdl.handle.net/11703/124619.

Scholz, Trebor. 2014. *Platform Cooperativism: Challenging the Corporate Sharing Economy*. New York: Rosa Luxemburg Stiftung.

Scholz, Trebor, and Nathan Schneider. 2016. *Ours to Hack and to Own: The Rise of Platform Cooperativism, a New Vision for the Future of Work and a Fairer Internet*. New York: OR Books.

Schragger, Richard. 2016. *City Power: Urban Governance in a Global Age*. Oxford, NY: Oxford University Press.

Scruggs, Gregory. 2018. Plaza Heralds New Era of Afrocentric Development in Seattle Neighborhood. *NextCity.org*. August 7, 2018. https://nextcity.org/daily/entry/plaza-heralds-new-era-of-afrocentric-development-in-seattle-neighborhood.

Selloni, Daniela. 2017. New Forms of Economies: Sharing Economy, Collaborative Consumption, Peer-to-Peer Economy. In *CoDesign for Public-Interest Services. Research for Development*. Cham, Switzerland: Springer.

Sen, Amartya. 1977. Rational Fools: A Critique of the Behavioral Foundations of Economic Theory. *Philosophy and Public Affairs* 6, no. 4: 317–344.

Sen, Amartya. 1992. *Inequality Reexamined*. Oxford: Clarendon Press.

Seoul Share Hub. 2016. Introducing Food Sharing Cases in Seoul. *Seoul Metropolitan Government—Seoul share Hub* (blog), June 23, 2016. http://sharehub.kr/share storyEn/news_view.do;jsessionid=4D480EF28FF48FAE2A6A9A260FBF577A?story Seq=87.

Shur-Ofry, Michal, and Ofer Malcai. 2019. Collective Action and Social Contagion: Community Gardens as a Case Study. *Regulation & Governance* (June). https://doi.org/10.1111/rego.12256.

Smith, Harry, and Tony Hernandez. 2020. Take a Stand, Own the Land: Dudley Neighbors, Inc., a Community Land Trust in Boston, Massachusetts. In *On Common Ground: International Perspectives on The Community Land Trust*, edited by John Emmeus Davis, Line Algoed, and María E. Hernández-Torrales, 283–294. Madison, WI: Terra Nostra Press.

Sobral, Laura. 2018. *Doing It Together—Cooperation Tools for the City Co-governance*. Berlin: Acidade Press.

Sørensen, Eva, and Jacob Torfing. 2011. Enhancing Collaborative Innovation in the Public Sector. *Administration & Society* 43, no. 8 (November): 842–868. https://doi:10.1177/0095399711418768.

Sorkin, David, and Wilhelm Von Humboldt. 1983. The Theory and Practice of Self-Formation (Bildung), 1791–1810. *Journal of the History of Ideas* 44, no. 1 (January–March): 55–73.

Sovacool, Benjamin K. 2013. Expanding renewable energy access with pro-poor public private partnerships in the developing world. *Energy Strategy Reviews* 1, no. 3: 181–192. https://doi.org/10.1016/j.esr.2012.11.003.

Spivack, Caroline. 2019. Vote for How City Funding Should Be Spent in Your Neighborhood. *Curbed NewYork* (blog), Apr 1, 2019. https://ny.curbed.com/2019/4/1/18290519/participatory-budgeting-vote-city-council-funding-neighborhood.

Stavrides, Stavros. 2016. *Common Space: The City as Commons*. London: Zed Books.

Steen, Kris, and Ellen van Bueren 2017. Defining Characteristics of Urban Living Labs. *Technology Innovation Management Review* 7, no. 7 (July): 21–33. https://timreview.ca/article/1088.

Sundararajan, Arun. 2016. The Sharing Economy. In *The End of Employment and the Rise of Crowd-Based Capitalism*. Cambridge, MA: MIT Press.

Swann, R. S., Shimon Gottschalk, Erick S. Hansch, and Edward Webster. 1972. *The Community Land Trust: A Guide to a New Model for Land Tenure in America*. Cambridge, MA: Center for Economic Development.

Sylvain, Olivier. 2012. Broadband Localism. *Ohio State Law Journal* 73, no. 4: 795–838. http://ir.lawnet.fordham.edu/faculty_scholarship/611.

Sylvain, Olivier. 2016. Network Equality. *Hastings Law Journal* 67: 443–498. http://www.hastingslawjournal.org/wp-content/uploads/Sylvain-67.2.pdf.

Tang, Belibei. 2015. Not Rural but Not Urban: Community Governance in China's Urban Villages. *China Quarterly* 223 (September) 724–744. https://doi.org/10.1017/S0305741015000843.

Taylor, Dorceta E. 2009. *The Environment and the People in American Cities, 1600s–1900s: Disorder, Inequality and Social Change.* Durham, NC: Duke University Press.

Thompson, Matthew. 2019. Playing with the Rules of the Game: Social Innovation for Urban Transformation. *International Journal of Urban and Regional Research* 46: 1168–1192.

Thorspe, Adam, and Sarah Rhode. 2018. The Public Collaboration Lab—Infrastructuring Redundancy with Communities in Place. *SheJi: The Journal of Design, Economics and Innovation* 4, no. 1: 60–74.

Tormos-Aponte, Fernando, and Gustavo A. García-López. 2018. Polycentric Struggles: The Experience of the Global Climate Justice Movement. *Environmental Policy and Government* 28, no. 4 (July-October): 284–294. https://doi.org/10.1002/eet.1815.

Townsend, Anthony M. 2014. *Smart Cities: Big Data, Civic Hackers, and the Quest for a New Utopia*, New York: W. W. Norton.

Townsend, Anthony M., Rachel Maguire, Mike Leibhold, and Mathias Crawford. 2010. *A Planet of Civic Laboratories: The Future of Cities, Information and Inclusion.* Palo Alto, CA: Institute for the Future & Rockefeller Foundation.

Tréguer, Félix, and Primavera De Filippi, Primavera. 2015. Wireless Community Networks: Towards a Public Policy for the Network Commons. In *Net Neutrality Compendium: Human Rights, Free Competition and the Future of the Internet*, edited by Luca Belli and Primavera De Filippi. New York: Springer.

Tribone, Lizzie. 2020. These Mothers Are Fighting for Their Families by Occupying Vacant Homes, *Rewire News Group* (blog), July 30, 2020. https://rewirenewsgroup.com/article/2020/07/30/these-mothers-are-fighting-for-their-families-by-occupying-vacant-homes/.

Turolla, Gregorio. 2020. Covid-19 Emergency in Naples: The Key Role of Self-Managed Spaces and Urban Commons. *URBACT* (blog), June 06, 2020. https://urbact.eu/covid-19-emergency-naples-key-role-self-managed-spaces-and-urban-commons.

Tysiachniouk, Maria S., and Andrey N. Petrov. 2018. Benefit sharing in the Arctic energy sector: Perspectives on corporate policies and practices in Northern Russia and Alaska. *Energy Research & Social Science* 39: 29–34. https://doi.org/10.1016/j.erss.2017.10.014.

UCLG Committee on Social Inclusion, Participatory Democracy and Human Rights. 2018. Defending a Policy for Urban Commons at the Local Level (Naples). *United*

Cities and Local Policies (UCLG) (blog), August 13, 2018. https://www.uclg-cisdp.org/en/news/latest-news/defending-policy-urban-commons-local-level-naples.

U.N.G.A. New Urban Agenda, 71st Sess., 68th plen. mtg., U.N. Doc. A/71/L.23 (Dec. 23, 2016). https://www.un.org/en/development/desa/population/migration/generalassembly/docs/globalcompact/A_RES_71_256.pdf.

UN-Habitat. 2016. *World Cities Report 2016: Urbanization and development. Emerging futures.* Nairobi, Kenya: UN-Habitat.

UN-Habitat. 2020. *World Cities Report 2020: Statistical Annex. Table D.1: Gross Domestic Product (GDP) & Gini Coefficient in Selected Cities.* Nairobi, Kenya: UN-Habitat. https://unhabitat.org/sites/default/files/2020/10/wcr_2020_statistical_annex_2.pdf.

United Nations, Department of Economic and Social Affairs, Population Division (UN DESA). 2019. *World Urbanization Prospects*: The 2018 Revision. New York: United Nations.

Unnikrishnan, Hita, B. Manjunatha, and Harini Nagendra. 2016. Contested Urban Commons: Mapping the Transition of a Lake to a Sports Stadium in Bangalore. *International Journal of the Commons* 10 (1):265–293. http://doi.org/10.18352/ijc.616.

Unnikrishnan, Hita, and Harini Nagendra. 2015. Privatizing the Commons: Impact on Ecosystem Services in Bangalore's Lakes. *Urban Ecosystem* 18, no. 2 (August): 613–632. https://doi.org/10.1007/s11252-014-0401-0.

URBACT. 2018. *Urbact Network—Refill.* URBACT (website). Accessed August 17, 2020. https://urbact.eu/Refill.

Verrucci, E., G. Perez-Fuentes, and T. Rossetto. 2016. Digital Engagement Methods for Earthquake and Fire Preparedness: A Review. *Natural Hazards* 83 (June): 1583–1604.

Wall, Elisabeth, and Remi Pelon. 2011. Sharing Mining Benefits in Developing Countries: The Experience with Foundations, Trusts and Funds. The World Ban. Extractive Industries and Development series. Washington, DC. https://openknowledge.worldbank.org/handle/10986/18290.

Wampler, Brian. 2000. A Guide to Participatory Budgeting. International Budget Partnership. https://www.internationalbudget.org/wp-content/uploads/A-Guide-to-Participatory-Budgeting.pdf.

Wang, Ruoniu, Claire Cahen, Arthur Acolin, and Rebecca J. Walter. 2019. Tracking Growth and Evaluating Performance of Shared Equity Homeownership Programs During Housing Market Fluctuations. Working Paper, Lincoln Institute of Land Policy. https://www.lincolninst.edu/publications/working-papers/tracking-growth-evaluating-performance-shared-equity-homeownership.

Weick, Karl. 1976. Educational Organizations as Loosely Coupled Systems. *Administrative Science Quarterly* 21, no. 1 (March): 1–19. https://doi.org/10.2307/2391875.

Weinberg, Bruce D., George R. Milne, Yana G. Andonova, and Fatima M. Hajjat. 2015. Internet of Things: Convenience vs. Privacy and Secrecy. *Business Horizons* 58, 6: 615–624. https://doi.org/10.016/j.bushor.2015.06.005.

Williamson, Theresa. 2018. Community Land Trusts in Rio's Favelas: Could Community Land Trusts in Informal Settlements Help Solve the World's Affordable Housing Crisis? *Land Lines Magazine* 30, no. 3 (July): 8–23.

Wilson, Douglas, Jesper Raakjær, Clyde Nielsen, and Paul Degnbol, eds. 2003. *The Fisheries Co-management Experience*. Dordrecht, Netherlands: Kluwer Academic.

Wolf-Powers, Laura. 2010. Community Benefits Agreements and Local Government: A Review of Recent Evidence. *Journal of the American Planning Association* 76, no. 2 (February): 141–159.

Won-Soon, Park. 2014. In Seoul, the Citizens Are the Mayor. *Public Administration Review* 74, no. 4 (July): 442–443. https://doi.org/10.1111/puar.12243.

Wood, Robert. 1958. The New Metropolis: Green Belts, Grass Roots or Gargantua. *American Political Science Review* 52 no. 1 (March): 108–122. https://doi.org/10.2307/1953016.

Woods, Melanie, Mara Balestrini, Sihana Bejtullahu, Stefano Bocconi, Gijs Boerwinkel, Marc Boonstra, et al. 2018. *Citizen Sensing: A Toolkit. Making Sense*. Dundee: University of Dundee. https://doi:10.20933/100001112.

World Habitat. 2015. Caño Martín Peña Community Land Trust. World Habitat Awards. Last accessed, September 8, 2020: https://www.world-habitat.org/world-habitat-awards/winners-and-finalists/cano-martin-pena-community-land-trust/.

Wu, Dan, and Sheila Foster. 2020. From Smart Cities to Co-Cities: Emerging Legal and Policy Responses to Urban Vacancy. *Fordham Urban Law Journal* 47, no. 4 (June): 909–939. https://ir.lawnet.fordham.edu/ulj/vol47/iss4/4.

Wu, Yan, and Wen Wang. 2012. Does Participatory Budgeting Improve the Legitimacy of the Local Government? A Comparative Case Study of Two Cities in China. *Australian Journal of Public Administration* 71, no. 2 (June):122–135.

Young, Iris Marion. 2000. *Inclusion and Democracy*. Oxford: Oxford University Press.

INDEX

Aarhus, Denmark, research dataset for, 244
Abandoned spaces. *See* Vacant land
Accra, Ghana
  limits of model in, 236
  research dataset for, 249
Acqua Bene Comune (ABC), 135
Adaptability, 207–210
Adelaide, Australia, research dataset for, 252
Administrative mediators, 31
Advisory boards/commissions, 165–166
Africa. *See also individual cities*
  cities coded and analyzed in, 249–250
  commons-based approaches in, 219–220
African American communities. *See* Black urban communities
Africatown Land Trust, 226
Agenda Tevere coalition, 228–229
Agents of Proximity, Bologna, 120
Agglomeration effects, 2–8, 32, 43–47
Agyeman, Julian, 11, 35
Alexander, Greg, 18
Alexander, Lisa, 92

Alternatives, space of, 135
Amsterdam, Netherlands
  research dataset for, 244
  Room for Initiative program, 143
  smart city technology in, 3, 12, 143–144, 207–208
Architects, neighborhood, 175–176
Arnstein, Sherry, 123, 167, 194
Art Commission of the City of New York, 76
Artificial intelligence (AI). *See* Smart cities
Ashdod, Israel, research dataset for, 250
Asia
  cities coded and analyzed in, 250–251
  commons-based approaches in, 219–220
Asilo Filangieri urban commons, 135, 136–137
Assets of Community Value (ACV), 133
Association Mastro Pilastro, 121
Athens, Greece, research dataset for, 244
Atlanta Land Trust, 99
AWMN wireless network, 213

Bamako, Mali, research dataset for, 249
Bandung, Indonesia, research dataset for, 250
Bangalore, India
  governance of urban lakes in, 70–71
  research dataset for, 250
Banjarmasin, Indonesia, research dataset for, 251
Barangay, Philippines, research dataset for, 251
Barcelona, Spain
  agglomeration effects in, 7
  Citizen Assets program in, 23, 140–142
  civic management in, 132, 139–142, 194, 204
  rebel city platform in, 17
  research dataset for, 244
  right to the city in, 3, 108
  technological infrastructure in, 215, 217
Barcelona en Comú, 17
Bari, Italy, research dataset for, 245
Baton Rouge, Louisiana, Co-City Baton Rouge project in, 222–227
Battipaglia, Italy, research dataset for, 245
Beirut, Lebanon, research dataset for, 251
Belgrade, Serbia, research dataset for, 245
Bella Fuori 3 pact, 124–125
Benefit-sharing agreements, 171–173
Beni comuni (common goods), 80–84
Benkler, Yochai, 53, 193
Bergrivier, South Africa, research dataset for, 250
Berlin, Germany
  digital stewardship in, 51, 213
  research dataset for, 245
Bilbao, Spain, research dataset for, 245
Bin-Zib community, Seoul, 87–89
Birmingham, Alabama, research dataset for, 245

Black and Brown Workers Cooperative, 40
Black Lives Matter (BLM) movement, 222, 234
Blackmar, Elizabeth, 44
Black urban communities. *See also* Inequalities; Wealth and resource disparities
  Co-City Baton Rouge project and, 222–227
  displacement of, 8, 14–17, 37–40, 86–87, 223
  environmental justice concerns in, 35
Bloc Bully IT Solutions, 179
Blockchain technologies, 180, 233
Blomley, Nicholas, 14, 18
Bloomberg Philanthropy Engaged Cities Award, 116
Bollier, David, 84, 205
Bologna, Italy
  Cities as a Commons project in, 116
  civic engagement and political culture of, 116–117
  Co-Bologna process in, 22–23, 115–126, 201
  collaboration pacts in, 111–112
  Engaged Cities Award given to, 116
  Iperbole digital platform in, 215
  Office for Civic Imagination in, 209
  research dataset for, 245
  urban co-governance in, 194
  urban labs in, 210
Bologna Regulation, 22–23
  adoption of, 115–116
  evaluation of, 123–126
  policymaking process for, 117–119
Borough councils, 164–165
Boston, Massachusetts
  agglomeration effects in, 7
  Dudley Street Neighbors Initiative in, 49–50, 94–95
Bottom-up urban commons, 28–29, 40–41, 80–84

Brazil
  community land trusts in, 221–222
  "right to the city" concept in, 3, 15–16
Bristol, England, research dataset for, 245
Broadband localism, 213
Broadband networks, 27, 51, 213–214
Brooklyn Community Board 16, 178
Brooklyn Public Library, 179
Brownsville neighborhood, New York City
  Brownsville Community Justice Center, 178
  Brownsville Houses Tenants Association, 178
  Brownsville Multi-Service Center, 178
  Brownsville Partnership, 178
  NYCx Co-Labs in, 178–180
Brussels, Belgium, research dataset for, 245
Budapest, Hungary, research dataset for, 245
Budgeting, participatory. *See* Participatory budgeting (PB)
Build Baton Rouge (BBR), 222–227
Business improvement districts (BIDs), 45, 77–80, 157

Caldeira, Teresa P. R., 52
Callan, Ireland, research dataset for, 245
Caltrans, 41
Cameroon, commons-based approaches in, 220
Can Batlló Self-Managed Community and Neighborhood Space Association, 142
Caño Martín Peña community land trust, 95–96
Cape Town, South Africa
  commons-based approaches in, 220
  research dataset for, 250
Capitalism, crowd-based, 204

Casablanca, Morocco, research dataset for, 250
Casa Ozanam Community hub pact (Turin), 184–185
Cascina Roccafranca Foundation, 184
Caserta, Italy, research dataset for, 245
Centocelle Archeological Park, 232
Central and Latin America, cities coded and analyzed in, 248–249. *See also individual cities*
Central Park, New York City, 44, 75–76
Central Park Conservancy, 75–76
Central Park Task Force, 75–76
Centro Antartide, 116
Challenge, right to. *See* Community right to challenge (CRTC)
Change, resistance to, 234–235
Charter of Common Rome, 17
Cheap talking phase, co-city protocol, 211
Chengdu, China, research dataset for, 251
Chicago, Illinois
  agglomeration effects in, 7
  community gardens in, 86
  land value in, 37
  NeighborSpace land trust in, 97–99
  participatory budgeting in, 169–170
  urban renewal programs in, 33
Chicago Park District, 97
Chief Science Offices, 208
Chief Technology Office/Officer (CTO), 207–208
Christchurch, New Zealand, research dataset for, 252
Chronic street nuisances, 45
Citidans, decision-making role of, 15
Cities as a Commons project, Bologna, 116
Citizen Assets program, Barcelona, 23, 140–142
Citizen Heritage Board, Barcelona, 140

Citizen Heritage of Community Use and Management, Barcelona, 140
Citizen participation, 164–173
  benefit-sharing agreements, 171–173
  community benefit agreements, 166–167
  community boards, 164–166
  ladder of, 123, 167–168
  neighborhood or borough councils, 164–165
  participatory budgeting, 30, 112, 119–120, 122, 129, 167–171, 215
Citizen Participation Regulations, Barcelona, 141
City-making, 106
City policies. See Public policies
City Science Offices (CSOs), 176
City Statute of Brazil, "right to the city" concept in, 3, 15–16
Ciudad Propuesta CDMX, 122
Civic and Collective Urban Uses, Turin, 145
Civic centers, 140
Civic Collective Management, Turin, 145
Civic deals, 145, 184–185
Civic eState network, 130
Civic flourishing environments, public spaces as, 136, 138
Civic management, declaratory approach tor, 132–144
Civic tech jobs, 217
Civic uses, 132–144
  Amsterdam Smart City program, 143–144
  Barcelona Citizen Assets program, 140–142
  characteristics of, 107–111, 132–133
  community right to bid, 133
  community right to challenge, 133–134
  Naples resolutions, 134–139
Civil Rights Movement, 222

Co-Bologna process, 115–126
  Agents of Proximity, 120
  Bologna Regulation, 22–23, 115–119, 123–126, 201
  as catalyst for other cities, 111–112
  Collaborare è Bologna process, 119, 215
  collaboration pacts in, 124–125
  Comunità Iberbole Platform, 119
  critique of, 126, 201
  evaluation of, 123–126
  Foundation for Urban Innovation, 119–120
  Incredibol policy, 119
  Office of Civic Imagination, 120
  participatory budgeting in, 119–120
  Pilastro neighborhood project, 119–122, 201, 209–210
  urban labs/collaboratories in, 119, 121–122
Co-Cities Guidance Codebook, 243–244
Co-City Baton Rouge (CCBR) project, 222–227
Co-city design principles. See Design principles
Co-city protocol
  cheap talking phase, 211
  mapping phase, 211
  modeling phase, 212
  practicing phase, 211–212
  prototyping phase, 212
  testing phase, 212
Co-City Turin project, 144–147, 181–185
Code for the City project, Mexico City, 122
Codigo para la Ciudad project, 122
Coding of cities, 243–244
Co-governance. See Urban co-governance
Cohesion Policy framework (EU), 190
Co-housing, 25, 28
  pooling and, 68, 204
  in Seoul, 87–89, 220

Co-Lab Challenges, NYCx Co-Labs, 179
Co-living, pact of, 209
Collaborare è Bologna process, Bologna, 119, 215
Collaboration pacts
  Co-Bologna process, 111–112, 115–126
  costs and challenges of, 108–110
  defined, 111
  evaluation of, 123–126
  New Turin Regulation, 145–146
  power of, 110–111
  Seoul Sharing city policy, 111–115
Collaborative economy, 53, 181
Collaborative governance, 194
  defined, 195–196
  failed practices, 105
  model of, 104–105
  regulatory or public policy approaches to. *See* public policies
Collaboratories. *See* Urban labs/collaboratories
Collective governance, 62–63, 193. *See also* Polycentric urban governance; Urban co-governance
  challenges of, 31
  collaboration pacts and, 124
  of community gardens, 72–73
  design principles for, 25, 28, 62–63
  future of, 221
  in Latin America, 221
  in NeighborSpace, 98–99
  pooling and, 52–56
  quintuple helix approach to, 30, 149–157, 183, 194
  recognition and respect for, 84
  regulatory or public policy approaches to. *See* public policies
  state enabling role in, 18, 71–72, 132
  of urban resources, 65–66
Collective political power, right to, 15
Colombes, France, research dataset for, 245

Co-management, 22, 24
  in Barcelona Citizen Assets program, 140–141
  pacts of collaboration and, 124, 127–128, 131
  stewardship models, 93
  in tech justice design principle, 216–217
Comedy of the commons, 47–48
Commoning, 19, 84–89
  defined, 84–85
  dynamics of, 85–89
Commoning.city web platform, 192, 239–240
Common interest communities, 92
Common pool resources, 20, 28, 43, 68. *See also* Governance
Commons, city as, 19–26. *See also* Urban commons
  concept of, 41–47
  cross-disciplinary application of, 20–22
  defined, 19
  exchange versus use value of land, 16
  historical and intellectual lineage of, 19–20
  Italian commons movement, 228
  regulatory or public policy approaches to. *See* public policies
  urban pooling economies, 52–56
Commons-based peer production (CBPP), 202
Commons Foundation, Turin, 204
Commons Transition Plan, Ghent, 130
Community Association for the Public Park of Centocelle, 230
Community Association of Neighborhood shops of Bolognina, 210
Community-based public-private partnerships (CBP3s), 158–159
Community benefit agreements (CBAs), 166–167
Community boards, 164–166

Community cooperative model, 189–190
Community cooperatives, 196, 231
Community EcoPark, Baton Rouge, 225
Community enablement
  neighborhood architects, 175–176
  technological tools for, 180
  urban labs or collaboratories, 55–56, 174–180, 207–208, 210, 217
Community gardens
  collective governance of, 72–73
  conflicts between residents and city over, 33–36
  creation of, 86
  pooling economy and, 203–204
Community improvement districts (CIDs), 77
Community Interest Companies (CICs), 189–190
Community land bank and trust (CLBT), 225–226
Community land trusts (CLTs), 28, 90–96
  Africatown Land Trust, 226
  Atlanta Land Trust, 99
  Caño Martín Peña, 95–96
  Co-City Baton Rouge project, 225–226
  construction of, 49–51
  Dudley Street Neighbors Initiative, 49–50, 94–95, 97
  governance structure of, 92–96, 196
  in Latin America, 50, 221–222
  legal structure of, 90–92
  motivations for, 50–51
  NeighborSpace, 97–99
  Oakland Community Land Trust, 41
  scaling of, 96–101
  in urban pooling economies, 54–55, 204
Community right to bid (CRTB), 133
Community right to challenge (CRTC), 133–134, 143
Community wireless mesh networks (CWN), 213–214
  construction of, 27, 51
  Coviolo Wireless project, 186
  pooling economy and, 203–204
Company-centered social responsibility (CCSR), 172
Comunità Iberbole Platform, Bologna, 119
Confagricoltura, 232
Conflict resolution, 59, 63
Constitutive policies, 107–111
  characteristics of, 29–30, 107–108
  Co-Bologna process, 111–112, 115–126
  costs and challenges of, 108–110
  declaratory policies blended with, 144–147
  Ghent's constitutive approach, 127–132
  Madrid Ordinance, 127–132
  power of, 110–111
  regulatory race toward the commons, 126–132
  Seoul Sharing city policy, 111–115
Constructed urban commons, 31.
  See also Community land trusts (CLTs); Limited-equity cooperatives (LECs); Urban commons; Wireless mesh networks
  characteristics of, 28–29, 62
  concept of, 41–47
  defined, 48
  examples of, 50–51
  Ostrom's framework applied to, 65–68
CooperACTiva, 228, 231
Cooperation, technology as driver of, 212
Cooperative governance. *See* Polycentric urban governance
Cooperatives, 189–190, 231
Coopify, 203
Co-ownership, 26, 176, 212, 215–217, 219

INDEX

Copenhagen, Denmark, research dataset for, 245
Co-production, 12, 18, 29. *See also* Public policies
Co-Roma project, 227–234
  CooperACTiva, 231
  GrInn.City, 231–233
  GrInn Lab, 229–231
  origins of, 227–228
  Rome Open Labs, 233–234
  smart co-district in, 229–231
  tech justice in, 216
  Tiber River, care and regeneration of, 229
Coruna, Spain, research dataset for, 245
COVID-19 pandemic, 4, 40, 115
Coviolo Wireless project, 186, 210
Creative class, 5–7
Creative use of law, 137
Crowd-based capitalism, 204
Cucina del Borgo pact, 185
Cultural institutions, role of, 7, 25, 80, 83, 236

Dahl, Robert, 200
Dakar, Senegal, research dataset for, 250
Daniels, Brigham, 46
Data collection, 240–244
Data selection, 240–244
Data set, 240–242
Data trusts, 218
Decentralization, 58, 135, 157, 209
Decidim Barcelona, 215
Declaration of Community Connectivity, 214
Declarations of Civic and Collective Use, 137
Declaratory policies, 107–111
  Amsterdam Smart City program, 143–144
  Barcelona Citizen Assets program, 140–142
  characteristics of, 29–30, 107–108, 132–133
  community right to bid, 133
  community right to challenge, 133–134, 143
  constitutive approach blended with, 144–147
  costs and challenges of, 108–110
  Naples resolutions, 134–139
  power of, 110–111
Deed-restricted homes, 90, 91, 93
DeFilippis, James, 100
de Magistris, Luigi, 135
Democratic deliberative property, 100
Design principles, 25–26, 31–32
  challenges of, 32
  co-governance, 25, 193–198
  enabling state, 25–26, 198–201
  Ostrom's framework for, 62–63, 192
  overview of, 191–192
  pooling economies, 26, 201–205
  tech justice, 26, 212–218
  urban experimentalism, 26, 205–212
Detroit, Michigan
  access to urban infrastructure in, 27
  agglomeration effects in, 7
  Creative Corridor Center, 7
  digital stewardship in, 51
  displacement of Black residents in, 37–40
  land bank program in, 39–41
  land value in, 38
  vacant or abandoned land in, 48
  wireless mesh networks in, 213
Developing economies, applications of co-city approach to, 236
Development Neighborhoods plans, Amsterdam, 143
Digital infrastructure, 114, 180, 212–218, 223
Digital stewardship, 51
di Robilant, Anna, 100, 202
Distributive justice, 17–18, 91

Dream Big Innovation Center, The, 179
Dublin, Ireland
  legal and property adaptation in, 90
  research dataset for, 245
Dudley Neighbors, Inc., 94
Dudley Street Neighbors Initiative, 49–50, 94–95, 97

Ecologically sustainable communities, 35
Economic production
  agglomeration effects and, 4–8
  cities as centers of, 1–2
Edge-cloud broadband networks, 213–214
Edinburgh, Scotland, research dataset for, 245
Education institutions, role of, 36–237
Eindhoven, Netherlands, research dataset for, 245
Ela, Nate, 86
Ellickson, Robert, 45, 64
Enabling state. *See also* Legal and property adaptation; Polycentric urban governance; Pooling economies; Public policies
  concept of, 25–26, 198–201
  construction of urban commons and, 28–29, 71–80
  facilitator role of, 18, 27, 107
  interventionist role of, 114–115
  research dataset for, 244–254
  resurgence of, 199
  role in polycentric governance, 57–60
Endogenous variables, 69
Energy microgrids, 51
Engaged Cities Award, 116
Environmental, social, and governance (ESG) approach, 186, 235
Environmental justice, 35, 51
Environmental sustainability, smart cities and, 10
Equitable autonomy, 202
Erasmus+EU program, 232

European Groups of Territorial Cooperation, 190
European Union (EU). *See also individual cities*
  Civic eState network, 130
  Cohesion Policy framework, 190
  funding programs of, 132, 183, 190
  regions/cities coded and analyzed in, 244–247
  Taxonomy Regulation, 235
European Urban Initiative, 190
Evaluative methodology, 206–207
Exchange value, 16, 27, 37–41
Exclusionary megacities, 37
Exogenous variables, 69
Experimentalism, 13
  adaptability in, 207–210
  city and citizen science in, 155
  in Co-Bologna process, 122
  defined, 26
  evaluative methodology of, 206–207
  importance of, 205–206
  iterative process of, 210–211
  legal and property adaptation, 28, 89–96
  regions/cities coded and analyzed for, 244–254
  role of, 205–212
Expulsions, 8, 16, 17–18
External mutualism, 236

Fab Labs, 220
Facilitator, state as, 18, 27, 107
Falklab project (Turin), 185
Faro Convention, 228
Favelas, 50, 221
Federcasa, 233
Fee simple, 91, 93
Fennell, Lee, 43
Fidenza, Italy, research dataset for, 245
Financing, 186–190
  Amsterdam Smart City program, 143
  commons foundations, 186–188

fundraising and urban commons finance tools, 188
  sustainable, 186, 189–190
First Life platform, 180
Fisheries projects, self-governance of, 64
5P partnerships. *See* Public-private-science-social-community partnerships (5P)
Florida, Richard, 5–7
Flynn, Ray, 94
Fondazione del Monte di Bologna, 116
Forest Preserve District of Cook County, 97
Forever Ultras pact, 125
Foundation for Urban Innovation, 119–120
Framework Regulation on Civic Participation, Madrid, 129
Free peering and transit, 214
Freifunk wireless network, 213
Friends of Park [X] groups, 74, 178
Frishmann, Brett, 193
Frug, Gerald, 106
Fundraising, 137, 186–190, 204
Future applications of co-city approach, 234–239
  developing economies, 236
  intergenerational alliances, 237–238
  legal and policy innovation, 235–236
  private sector investments, 235
  resistance to, 234–235
  role of scientific, knowledge, education, and cultural institutions in, 36–237
Future generations, rights of, 237

Gdansk, Poland, research dataset for, 245
General interest, 119, 134, 138, 162, 189, 194, 200
Gentrification
  community land trusts as response to, 50, 99, 221
  displacement by, 37–41, 82, 89–90
  privatization and, 78–79
Ghana Institute for Public Management, 236
Ghent, Belgium
  constitutive approach of, 127–132
  research dataset for, 245
Glaeser, Edward, 5
Glasgow, Scotland, research dataset for, 245
Global South, 236
  applications of co-city approach to, 24, 236
  peripheral urbanization in, 52
  rural to urban migration in, 2
  urban interaction space in, 48
Goodman, Ellen, 11
Gothenburg, Sweden, research dataset for, 245
Governance. *See also* Collective governance; Polycentric urban governance; Urban co-governance
  of community land trusts, 92–96
  conflict resolution in, 63
  failure of, 11
  of green urban commons, 69–71
  of limited-equity cooperatives, 92–96
  megacity model of, 106–107
  of natural resource commons, 62–63
  nested, 96–101
  Ostrom's design principles for, 41–43, 46, 56–58, 62–63, 193, 195, 197–198
  property stewardship models of, 92–96
  right to, 15
  self-governance, 63–64
  traditional tripartite structure of, 91, 92, 99
  of urban green commons, 69–71
Governance property, 91
Graffiti, 45, 79, 124
Great Migration, 39

Green and sustainable finance, 186, 189–190
GreenThumb Program, New York City, 73
Green urban commons, governance of, 69–71
Grenoble, France, research dataset for, 245
GrInn.City, 231–233
GrInn Lab, 229–231
Gross domestic product (GDP), contribution of cities to, 1–2
Guangzhou, China
  research dataset for, 251
  urban villages in, 220–221
Guifi wireless network, 213

H2020, 190
Habitat project, Turin, 185
Hackney Community Transport, 189–190
Hamburg, Germany, research dataset for, 245
Hardin, Garret, 42–44, 67
Harlem, New York City, 213–214, 223
Harvey, David, 14, 16, 47
Helfrich, Silke, 84, 205
Helsinki, Finland, research dataset for, 245
Hess, Charlotte, 193
High human capital individuals, 5. *See also* Agglomeration effects
High Line (New York City), 47
Hollands, Robert, 9
Holon, Israel, research dataset for, 251
Homogeneous communities, self-governance in, 64
Hong Kong, research dataset for, 251
Horizon 2020 OpenHeritage EU project, 228
"House of Emerging Technologies" program (Rome), 233–234
Housing for All (Iaione et al.), 233

Housing initiatives
  Bin-Zib co-housing communities, 87–89
  Caño Martín Peña community land trust, 95–96
  co-housing, 25, 28, 68, 87–89, 204, 220
  deed-restricted homes, 90, 91, 93
  Dudley Street Neighbors Initiative, 49–50, 94–95
  GrInn.City, 231–233
  limited-equity cooperatives, 28, 86–87, 90–96
  occupy movement and, 82–84
  HUB Underground Base pact, 125
  Huron, Amanda, 86–87
Hypervacancy, 48–49

Iasi, Romania, research dataset for, 245
Ibeju Lekki Local Government, benefit sharing arrangement with, 172–173
Idea collection maps, 169
Incredibol policy, Bologna, 119
Inequalities. *See also* Minority communities; Urban poor
  in collective pacts, 206
  in environmental justice, 35
  governance and, 31, 59
  nested institutions and, 78
  in participatory budgeting, 170–171
  in smart cities, 216
  wealth and resource disparities, 2–3, 7–8, 35, 158, 212–218
Institutional analysis and development (IAD) framework, 69–70
Institutional spaces, 31, 109, 116, 174, 207
Intergenerational alliances, 237–238
Internet Governance Forum (IGF)
  Dynamic Coalition on Community Connectivity (DC3), 214
Interventionist state, 114–115
Iperbole digital platform, 215

Irrigation projects, self-governance of, 64
Istanbul, Turkey, peripheral urbanization in, 52
Italian Constitution, rights of future generations in, 237
Iterative process, 210–211

Jacobs, Jane, 5, 36
Jerusalem, research dataset for, 251
Jim Crow era, 222
Johannesburg, South Africa, research dataset for, 250
Just sustainability, 12, 35

Karachi, Pakistan, research dataset for, 251
Kathamandu, Nepal, research dataset for, 251
Katyal, Sonia K., 83
Kenya, commons-based approaches in, 220
Ker Thiossane project, 220
Keyboard, video, and mouse (KVM) devices, 214
Kigali, Rwanda, research dataset for, 250
Kinshasa, DR Congo, research dataset for, 250
Kitchin, Dan, 10
Knowledge House, The, 179
Knowledge institutions, role of, 18, 222
  agglomeration effects and, 7
  in Co-Bologna process, 119
  in collaboration pacts, 111
  enabling state and, 199
  in future applications of co-city approach, 236–238
  in pooling, 53, 89
  public-private-science-social-community partnerships, 160–162, 172–173, 181
  in urban governance, 25, 30, 107, 110, 151–153
  in urban labs, 55

Koregaoni, India, research dataset for, 251
Kyoto, Japan, research dataset for, 251

Lab for the City, Mexico City, 55
LabGov, 227–234
Laboratorio para la ciudad (Mexico City), 55, 122, 177
Labs. *See* Urban labs/collaboratories
Labsus, 116
Ladder of citizen participation, 123, 167–168
Lagos, Nigeria
  benefit sharing arrangement in, 172–173
  research dataset for, 250
Lahore, India, research dataset for, 251
Lakes, governance of, 70–71
Land, urban. *See also* Community land trusts (CLTs); Urban commons; Vacant land
  bottom-up approaches to, 40–41
  community gardens, 33–36, 72–73, 86, 203–204
  conflicts between residents and city over, 33–41
  exchange versus use value of, 16
  privatization of, 37–38
  tension between exchange and use value of, 27, 37–41
  top-down approaches to, 39–40
  value of, 37–38
Land bank program, Detroit, 39–41
Landmarks Preservation Commission, 76
Land trusts. *See* Community land trusts (CLTs)
Last-mile network connectivity gap, 51
Lawanson, Taibat, 172–173
Lefebvre, Henry, 14, 15, 19

Legal and property adaptation, 28, 89–96, 235–236
　community land trusts, 41, 49–51, 54–55, 90–96, 196, 221–222, 225–226
　limited-equity cooperatives, 28, 86–87, 90–96
　as response to gentrification, 89–90
Legal hacking, 137
Lekki Worldwide Investment Limited, 172–173
Lepida, 186
Lille, France, research dataset for, 246
Limited-equity cooperatives (LECs), 28, 86–87, 90–96
　governance structure of, 91–96
　legal structure of, 90–91
　limited-equity homeownership, 93–94
Linebaugh, Peter, 18
Lisbon, Portugal, research dataset for, 246
Liverpool, England, research dataset for, 246
Local administrations, as enabling platforms, 22, 111, 120, 136–140, 146, 188, 200–201
Local Entities Assets Regulations, Barcelona, 141
Local environmental stewardship
　community gardens as, 33–36
　defined, 34
Logan, John, 37
Loitering, 45
Lomé, Togo, research dataset for, 250
London, England
　agglomeration effects in, 7
　economic output of, 1–2
　research dataset for, 246
Los Angeles, California
　community benefit agreements in, 166–167
　economic output of, 1
　land vacancy in, 41, 48
　land value in, 37

Lucca, Italy, research dataset for, 246
Luiss University LabGov, Co-Roma project and, 227–234
Lyon, France, research dataset for, 246

Made in Brownsville, 178
Madison, Michael, 193
Madrid, Spain
　co-governance in, 194
　Madrid Ordinance, 127–132
　research dataset for, 246
Magnet cities, vacancy rates in, 48
Malmo, Sweden, research dataset for, 246
Management assembly, 137
Mantova, Italy, research dataset for, 246
Mapping phase, co-city protocol, 211
Maribor, Slovenia, research dataset for, 246
Marseille, France, research dataset for, 246
Martin Peña Canal Special Planning District, 95–96
Massarosa, Italy, research dataset for, 246
Mataró, Spain, research dataset for, 246
Matching, 5
Matera, Italy, research dataset for, 246
Mayor's Fund to Advance New York City, 55
Mayor's Office of the Chief Technology Officer (MOCTO), 55, 177–178
Mazzucato, Mariana, 218
McLaren, Duncan, 11
Megacity model of governance, 106–107
Melbourne, Australia, research dataset for, 252
Member's Media, Ltd. Cooperative, 203
Mesh networks. *See* Community wireless mesh networks (CWN)
Messina, Sicily, research dataset for, 246
Mexico City
　Code for the City project, 122
　economic output of, 1–2

Laboratory for the City, 55, 122, 177
peripheral urbanization in, 52
Proposed City CDMX, 122
Right to the City Charter, 3, 16
Milan, Italy
  agglomeration effects in, 7
  occupy movement in, 83–84
  research dataset for, 246
Minority communities
  Co-City Baton Rouge project and, 222–227
  conservation of resources of, 172–173
  displacement of, 8, 14–17, 37–40, 86–87, 165, 223
  environmental justice concerns in, 35
  impact of privatization on, 78–79
Modeling phase, co-city protocol, 212
Molotoch, Harvey, 37
Mombasa, Kenya, research dataset for, 250
Moms4Housing, 40–41
Monocentric governments, 57
Montepellier, France, research dataset for, 246
Moor, Tine De, 193
Moses, Robert, 165
Moviemento Urbano Popular (MUP), 16
Multi-actor governance. *See* Polycentric urban governance
Mumbai, India, research dataset for, 251
Mumford, Lewis, 181
Municipal Body Regulations, Barcelona, 141
Municipal Charter of the City of Barcelona, 141
Mutualism, external, 236

Nagendra, Harini, 70
Nairobi, Kenya, research dataset for, 250
Nantes, France, research dataset for, 246
Naples, Italy
  co-governance in, 194
  declaratory policies in, 134–139

population and economic output of, 133–134
  rebel city platform in, 17
  research dataset for, 246
Narni, Italy, research dataset for, 246
National Agency for New Technologies, Energy and Sustainable Economic Development (Italy), 228
Natural resource commons, 20
  Ostrom's framework for, 28, 61–65
  urban commons compared to, 28, 61–62
Neco wireless network, 213
Neighborhood architects, 175–176
Neighborhood-as-a-commons program, 174–175
Neighborhood commons
  business improvement districts and, 45, 77–80, 157
  tragedy of the commons scenario, 45–46
Neighborhood councils, 164–165
Neighborhood Houses, Turin, 183–184
Neighborhood improvement districts (NIDs), 77
Neighborhood innovation labs. *See* Urban labs/collaboratories
Neighborhood pooling economies
  defined, 181
  Reggio Emilia projects, 185–186
  Turin Co-City and Co4Cities projects, 181–185
  urban co-governance as driver for, 181–186
NeighborSpace, 97–99
Neoliberalism, 158, 199
Nested enterprises, 46
  in Co-Bologna project, 125
  enablement of, 57–58, 60, 63, 78, 103
  NeighborSpace land trust, 97–99
  nested urban commons, 96–101
  urban co-governance and, 110, 123, 156

Networks
  broadband, 27, 51, 213–214
  collective ownership of, 214
  community wireless mesh, 27, 51, 186, 203–204, 213–214
  network equality, 214
  network neutrality, 214
New Turin Regulation, 144–147, 186–190
New urban crisis, 7–8
New York City
  access to urban infrastructure in, 27
  agglomeration effects in, 7
  Central Park, 75–76
  community boards in, 164–166
  community land trusts in, 54–55
  digital divide in, 213
  economic output of, 1
  GreenThumb Program, 73
  High Line, 47
  land value in, 37
  Mayor's Office of the Chief Technology Officer, 177–178
  NYCx Co-Lab initiative, 55–56, 177–180, 217
  participatory budgeting in, 167–171
  as smart city, 3, 213–214
  Times Square, 79
  tragedy of the commons scenario and, 44
  urban pooling economy in, 54–55
  urban renewal programs in, 33
  vacancy rates in, 48
Ninux wireless network, Germany, 213
Nobel Prize, 19, 41, 62
Nongovernmental organizations (NGOs)
  collaboration pacts with, 108, 116, 124–125
  co-management with, 140
Non-Profit Utilities (NPUs), 196
Nonrivalrous resources, 44
North America, cities coded and analyzed in, 247–248. *See also individual cities*

Nubian Square, Boston, 49–50
Nussbaum, Martha, 12
NYCx Co-Lab initiative, 55–56, 177–180, 217
NYCx Co-Labs, 177–180, 217

Oakland, California, land vacancy in, 40–41
Oakland Community Land Trust, 41
Observatory of Urban Commons and Participatory Democracy, Naples, 135
Occupy movement, 82–84
Occupy PHA, 40
Oceania, cities coded and analyzed in, 251–252. *See also individual cities*
Office for Civic Imagination, Bologna, 120, 209, 210
Olajide, Oluwafemi, 172–173
Open access resources, vacant land as, 80–84
Oslo, Norway, research dataset for, 246
Osteria di Centocelle (Rome), 230
Ostrava, Czech Republic, research dataset for, 246
Ostrom, Elinor
  collective and self-governance scenarios, 41–43, 46, 56–58, 193, 195
  design principles of, 62–63, 192
  institutional analysis and development framework, 69–70
  natural resource commons framework, 19–20, 28, 61–65
  Nobel Prize awarded to, 19, 41, 62
  polycentric governance scenarios, 56–57, 64–65, 193, 197–198
  social-ecological system framework, 70–71
  water basin management findings, 78
  work applied to urban commons, 41–42, 65–68
Ostrom, Vincent, 57
OUTakes archive, 124

Pact of co-living, 209
Pacts of collaboration. *See* Collaboration pacts
Padua, Italy, research dataset for, 246
Palermo, Sicily, research dataset for, 246
Panhandling, 45
Paris, France
  agglomeration effects in, 7
  economic output of, 1–2
  participatory budgeting in, 215
  research dataset for, 246
Park conservancies, 45, 75–77
Parks
  friends groups for, 74–75
  legal protection of, 35–36
  state enabling role in, 73–77
  tragedy of the commons scenario and, 43–47
Participant assembly problem, 43–44
Participation, technology as driver of, 212
Participatory budgeting (PB), 30
  in Co-Bologna process, 119–120
  in Ghent, 129
  in Mexico City, 112, 122
  motivations for, 167–171
  in New York City, 168–169
  in Paris, 215
Participatory policies, 30–31
Partnerships
  benefit-sharing agreements, 172
  business improvement districts, 77–80
  community improvement districts, 77
  Friends of Park [X] groups, 74–75
  neighborhood improvement districts, 77
  park conservancies, 75–77
  public-community, 31, 159–160, 191
  public-community-private, 191
  public-private, 28, 71–80, 157–158
  public-private-community, 31

public-private-science-social-community, 160–161, 172–173, 181
  unequal power dynamics in, 171
Partnerships for Parks, New York City, 74
Paternalistic benefit-sharing agreements, 172
Peer-to-peer economies, 181
Peñalver, Eduardo Moises, 83
Peniche, Portugal, research dataset for, 247
Peripheral urbanization, 52
Permanent Council, Turin, 144–145
Philadelphia, Pennsylvania, land vacancy in, 40–41, 48
Philadelphia Housing Authority (PHA), 40–41
Piantala pact, 125
Pilastro neighborhood project, 119–122, 201, 209–210
Pilastro Northeast Development Agency, 209
Pitkin Avenue BID, 178–179
Pittsburgh, Pennsylvania, agglomeration effects in, 7
Plank Road Corridor, Baton Rouge. *See* Co-City Baton Rouge (CCBR) project
Platform cooperativism, 203
Platform economy, 181
Policies. *See* Public policies
Policy innovation labs, 176
Polycentric urban governance, 29, 64–65, 193–194. *See also* Urban co-governance
  advantages of, 56–58, 191
  criticisms of, 58–59
  enabling of, 27, 56–60
  importance of, 196–198
  origins of, 196–197
  Ostrom's analysis of, 56–57, 64–65, 193, 197–198
  public entrepreneurships in, 57

Polycentric urban governance (cont.)
  public-private partnerships in, 29
  state role in, 57–60
Pooling economies
  concept of, 18–19, 52–56, 201–203
  defined, 26, 68, 89, 181
  design principles for, 201–203
  NeighborSpace initiative and, 97–99
  regions/cities coded and analyzed for, 244–254
  role of, 201–205
  Seoul's Sharing city policy and, 115
  state role in, 28
  types of, 203–205
  urban co-governance as driver for, 181–186
  urban commons and, 43, 68, 89, 97
  urban pools, 202
Pool resources, urban commons compared to, 27
Porto Alegre, Brazil, participatory budgeting in, 168, 170
Powles, Julia, 11
Practicing phase, co-city protocol, 211–212
Predatory city, Detroit as, 38–40
Presov, Slovakia, research dataset for, 247
Pretoria, South Africa, commons-based approaches in, 220
Privacy, network, 214
Private sector investments, 235
Privatization of land, 37–38, 78–79
Procurement, 134, 139, 141, 200
Property adaptation. *See* Legal and property adaptation
Property tax foreclosures, 37–41
Proposed City CDMX, Mexico City, 122
Prototyping phase, co-city protocol, 212
Proximity, *tragedy of the commons* scenario and, 43–47
Public actors, 104
Public Administration Assets Management, Spain, 128

Public Archeological Park of Centocelle, 230
Public calls, 119, 128
Public-citizen partnerships, 159–160
Public-community partnerships (PCPs), 31, 159–160, 191. *See also* Citizen participation; Community enablement
Public-community-private partnerships (PCPPs), 191
Public entrepreneurship, 57
Public participation. *See* Citizen participation
Public policies, 22–24, 29–30. *See also* Collaboration pacts; Constitutive policies; Declaratory policies
  Amsterdam Smart City program, 143–144
  Barcelona Citizen Assets program, 140–142
  blended approach to, 144–147
  civic uses and civic management, 132–144
  Co-Bologna process, 111–112, 115–126
  collaborative governance reflected in, 103–107
  community right to bid, 133
  community right to challenge, 133–134, 143
  costs and challenges of, 108–110
  creative use of law and, 137
  declaratory versus constitutive, 29–30, 107–111
  emergence of, 29–30, 103–107
  in future applications of co-city approach, 235–236
  Ghent's constitutive approach, 127–132
  legal and policy innovation, 235–236
  Madrid Ordinance, 127–132
  motivations for, 104–105
  Naples resolutions, 134–139

participatory, 30–31
power of, 110–111
regulatory race toward the commons, 126–132
Seoul Sharing city policy, 111–115
urban pooling economies, 52–56
urban renewal, 223
Public-private-community partnerships (PCPPs), 31
Public-private partnerships, 28
  enablement of urban commons by, 71–80
  urban revitalization and, 45–46
Public-private partnerships (PPPs), 157–158
Public-private-science-social-community partnerships (5P), 160–161, 172–173, 181
Public-private-science-social-community partnerships in, 173
Public procurement, 134, 139, 141, 200
Public-public partnerships, 159–160
Public sector, 104
Public trust doctrine, 35–36
Public utilities, collaborative governance of, 196
Pune, India, research dataset for, 251
Purcell, Mark, 14

Quayside Waterfront smart city project, 217–218
Quintuple helix governance, 30, 149–157, 183, 194

Racial justice, 4, 222–227, 234. *See also* Black urban communities; Inequalities
Rawls, John, 12
Rebel cities, 16–17
Rebel city platforms, 16–17
Reclaiming Our Homes, 41
Red Hook Wi-Fi, 51, 213

Reggio Emilia, Italy
  collaboratories in, 174–177
  Coviolo Wireless project, 186
  neighborhood-as-a-commons policy, 174–176, 210
  Regulation for Citizenship Agreement, 174
  research dataset for, 247
  wireless community network in, 204
Regies de quartier model, 120–121
Regional research datasets, 244–253
  Africa, 249–250
  Asia, 250–251
  Central and Latin America, 248–249
  Europe, 244–247
  North America, 247–248
  Oceania, 251–252
Regional water basin management, 78
Register of the Guarantors, Turin, 144
Regulation. *See* Public policies
Regulation (EU) 2020/852, 235
Regulation for Citizenship Agreement, 174
Regulation for Collaboration between Citizens and the City for the Care and Regeneration of the Urban Commons. *See* Bologna Regulation
Regulation on Local Entities Assets Management, Spain, 128
Research methodology, 240–244. *See also* Design principles
  case study source information, 240–242
  co-cities data set, 242–243
  coding of cities, 243–244
  Commoning.city web platform, 192, 239–240
  data selection and collection, 24–25, 240–244
  goals of, 25
  regions and cities coded and analyzed, 244–253
Resource pooling. *See* Pooling economies
Resurgence of state, 199

Revitalization plans, 6
Revolutionary Workers Collective, 40
"Right to the city" concept, 1-3, 13-19
  defined, 14
  Latin American and European contexts of, 15-16
  Mexico City's Right to the City Charter, 3, 16
  origins of, 13
  rebel cities and, 16-17
  uncertainties in, 14-15
Rio de Janeiro, Brazil, community land trusts in, 50
Risk prevention, 186
Rivalrous resources, 44
Roma Tre Information Engineering Department, 234
Rome, Italy
  occupy movement in, 82-83
  research dataset for, 247
  "right to the city" concept in, 17
  Valle Theatre in, 82
Rome Open Labs, 233-234
Room for Initiative program, Amsterdam, 143
Rose, Carol, 47, 193
Rosenzweig, Roy, 44
Rotterdam, Netherlands, research dataset for, 247

Safe and Thriving Nighttime Corridors challenge, 179
Safransky, Sara, 39-40
Saint Joseph Church, Ghent, 130, 132
Samenlevingsopbouw Gent, 129
Sandel, Michael, 12
San Donato Neighborhood administration, Bologna, 121
San Francisco, California
  agglomeration effects in, 7
  land value in, 37
  vacant land in, 48
San Giovanni Tower, 230

San Juan, Puerto Rico, community land trusts in, 50, 95-96, 222
San Tammaro, Italy, research dataset for, 247
Santiago, Chile, peripheral urbanization in, 52
Sao Paulo, Brazil, peripheral urbanization in, 52
Sapienza Innovation, 234
Sapienza University, 229
Sarantaporo, Greece, research dataset for, 247
Sarker, Ashutosh, 58
Sassari, Italy, research dataset for, 247
Sassen, Saskia, 17, 37
Scientific institutions, role of, 36-237
Seattle, Washington
  Africatown Land Trust in, 226
  community boards in, 165
  vacant land in, 48
Self-governance, 36, 47, 58, 134, 145-146, 193, 195-197. *See also* Polycentric urban governance; Urban co-governance
Sen, Amartya, 12
Seoul, South Korea
  Bin-Zib co-housing communities in, 87-89
  economic output of, 1-2
  interventionist state in, 114-115
  population and economic output of, 112
  research dataset for, 251
  Seoul Sharing city policy, 111-115
  smart city technology in, 12
  urban villages in, 220
Seoul Community Support Center (SCSC), 113-114, 220
Shared infrastructure. *See* Urban commons
Shared Management of the Ex Serre Giardini Margherita pact, 125

Shareholder benefit-sharing agreements, 172
Sharing economy, 181, 204
Sharing Economy International Advisory Board, Seoul, 113
Sharing Promotion Fund, Seoul, 113
Shenyang, China, research dataset for, 251
Shenzhen, China
  research dataset for, 251
  urban villages in, 220–221
Sidewalk Labs, 11, 217
Sidney, Australia, research dataset for, 252
Slum clearance programs, 36
Smart cities, 3, 8–13
  Amsterdam Smart City program, 143–144
  characteristics of, 9–10
  community enablement through, 180
  in Co-Roma project, 229–231
  digital divide, 213, 223
  potential of, 8–9
  Seoul Sharing city policy, 114
  sharing city paradigm for, 11–13
  tech justice design principle, 212–218
  tensions in, 10–11
Smart City program, Amsterdam, 143–144
Social and economic pooling. *See* Pooling economies
Social benefits, 143
Social contagion, 73
Social-ecological system (SES) framework, 70–71
Social influence, 73
Social justice. *See* Black urban communities; Environmental justice; Inequalities; Racial justice; Wealth and resource disparities
Social licenses, 172
Social management, 140–141, 214
Social norms, self-governance and, 64

Social purpose vehicles, 201
Social Taxonomy Report, 190
Soft infrastructure, 10
Solidarity economy, 47–48, 53
South Africa
  benefit-sharing agreements in, 171–172
  commons-based approaches in, 220
  occupy movement in, 82
Special water districts, 78
State, enabling. *See* Enabling state
Steering assembly, 137
Stewardship models, 34, 92–96
St. Louis, Missouri, agglomeration effects in, 7
Strandburg, Katherine, 193
Street Art Pact at the Zaccarelli Center, 124
Structural racism. *See* Black urban communities; Inequalities; Racial justice; Wealth and resource disparities
Suburbs
  wealth and resource disparities in, 8
  white flight to, **222**
Sustainability
  in Amsterdam Smart City program, 143
  ecological, 35
  environmental, 10
  just, 12, 35
  Seoul Sustainable Development Goals, 114
  social, 10
  sustainable finance framework, 186, 188–190
  sustainable investment, 235
Sylvain, Olivier, 213
SynAthina, 215

"Take back the land" movement, 82–84
Taxonomy Regulation, 235
Tech Boards, NYCx Co-Labs, 178–180

Tech justice
  dimensions of, 215–218
  examples of, 212–218
  research dataset for, 244–254
Technical Board, Turin, 145
Technological infrastructure, access to, 212–218
Technological sovereignty, 217
Technology-enabled smart cities. *See* Smart cities
Tel Aviv, Israel, economic output of, 1–2
Tender procedures, 128
Testing phase, co-city protocol, 212
Thematic tables, 137
Third sector, 59
3 Black Cats Cafe, 179
Three Ts of development, 6
Tiber for all (Tevere per Tutti) foundation, 229
Tiber River, care and regeneration of, 229
Times Square, New York City, 79
Tokyo
  economic output of, 1–2
  research dataset for, 251
Tolerance, 6
Toronto, Canada
  economic output of, 1–2
  Quayside Waterfront smart city project, 11, 217–218
Torre Spaccata, 232
Tor Vergata School of Engineering, 234
Tower of Centocelle (Rome), 230
Tragedy of the Commons, The (Hardin), 42–44, 67
Tragedy of the commons scenario, 42–47, 67–68
Tripartite governance, 91, 92, 99
Tunnel of Centocelle, 230
Turin, Italy
  Co-City and Co4Cities projects, 144–147, 181–185
  Commons Foundation, 204
  financing innovations in, 186–190

First Life platform, 180
  research dataset for, 247
Tuscany, Italy, citizen participation in, 116–117

Uganda, commons-based approaches in, 220
United Kingdom (UK)
  community right to bid, 133
  community right to challenge, 133–134
United Nations (UN), Internet Governance Forum Dynamic Coalition on Community Connectivity, 214
URBACT program, 132, 138, 143, 183, 190
Urban Center of Bologna, 119–120
Urban civic communities, 138
Urban Civic Use Regulation, 137
Urban civic uses, collaborative governance of, 196
Urban co-governance. *See also* Collective governance; Polycentric urban governance; Public policies; Urban labs/collaboratories
  benefit-sharing agreements, 172
  business improvement districts, 77–80
  citizen participation in, 164–173
  collaborative, 194, 195–196
  community enablement in, 174–180
  community improvement districts, 77
  conceptualization of, 30–31, 149–150, 193–198
  defined, 25, 194
  as driver for neighborhood pooling economies, 181–186
  financing of, 186–190
  levels/steps of, 193–195
  neighborhood improvement districts, 77
  origins and influences on, 193
  park conservancies, 75–77

participatory policies compared to, 30–31
public-community partnerships, 31, 159–160, 191
public-community-private partnerships, 191
public-private-community partnerships, 31
public-private partnerships, 28, 71–80, 157–158
public-private-science-social-community partnerships, 160–162, 172–173, 181
quintuple helix model of, 30, 149–157, 194
research dataset for, 244–254
shared, 194, 195
unequal power dynamics in, 171
Urban commons, 19–26. *See also* Community land trusts (CLTs); Pooling economies; Vacant land
adaptability, 207–210
bottom-up, 28–29, 40–41, 80–84
characteristics of, 61–62, 67–68
comedy of the commons scenarios, 47–48
common pool resources compared to, 28, 43–44
community gardens, 33–36, 72–73, 86, 203–204
constructed, 28–29, 47–52
creation of (commoning), 84–89
cross-disciplinary application of, 20–22
defined, 19
examples of, 49–52
exclusionary effects in, 46–47
governance property, 91
green, 69–71
historical and intellectual lineage of, 19–20
legal and property adaptation in. *See* legal and property adaptation
natural resource commons compared to, 61–62

neighborhood-as-a-commons program, 174–176
nested, 96–101
Ostrom's framework applied to, 41–42, 65–68
public-private partnerships for, 71–80
social interaction in, 47–48
solidarity increased by, 47–48
state enabling role in, 27, 43, 71–80
top-down, 28–29, 80–84
tragedy of the commons scenario, 42–47, 68
Urban Commons Foundation, Turin, 145, 146, 186–188
Urban creativity, 119, 138
Urban experimentalism, 13, 28, 192. *See also* Urban labs/collaboratories
city and citizen science in, 155
in Co-Bologna process, 122
defined, 26
evaluative methodology of, 206–207
importance of, 205–206
iterative process in, 210–211
organizational innovations in, 208–212
scaling of, 207–208
Urban frameworks, 4–8
overview of, 1–4
"right to the city" concept, 3, 13–19
technology-enabled smart cities, 3, 8–13
urban agglomeration, 3, 4–8
Urban governance. *See* Governance
Urban Innovative Actions initiative (UIA), 183, 190
Urbanization, 1, 20, 43, 52
in China, 220–221
collective power of residents over, 14
green urban commons and, 69, 71
peripheral, 52
property value and, 91
rebel cities, 16

Urbanization (cont.)
 "right to the city" concept, 1–3, 13–19
 smart cities as response to, 9
Urban labs/collaboratories, 55–56, 177–180, 207–208, 210
 in Amsterdam Smart City platform, 207–208
 in Co-Bologna project, 119, 121–122, 209–210
 defined, 174
 in Mexico City, 177
 in Mexico City, 55, 122, 177
 neighborhood architects for, 175–176
 NYCx Co-Labs, 55–56, 177–180, 217
 policy innovation labs compared to, 176
 in Reggio Emilia, 174–177
Urban land. *See* Land, urban
Urban peripheries, growth of, 2, 66
Urban pooling economies. *See* Pooling economies
Urban pools, 202
Urban poor
 displacement of, 2, 14–16, 89–90, 165
 impact of privatization on, 78–79
 marginalization of, 170, 180
 participatory budgeting and, 170–171
Urban Popular Movement, 16
Urban renewal era
 communities displaced by, 165, 223
 community boards in, 164–166
 failure of, 33
 slum clearance programs, 36
Urban revitalization, 6
 Co-City Baton Rouge project, 222–227
 communities displaced by, 223
 community gardens as, 33–36, 72–73, 86, 203–204
 public-private partnerships in, 45–46
Urban villages, 220–221
Use value, 16, 27, 37–41
Utrecht, Netherlands, research dataset for, 247

Vacant land
 amount of, 48–49
 bottom-up approaches to, 40–41
 community gardens created from, 33–36
 conflicts between residents and city over, 33–41, 80–84
 Dudley Street Neighbors Initiative, 49–50, 94–95, 97
 hypervacancy, 48–49
 NeighborSpace land trust, 96–101
 tension between exchange and use value of, 37–41
 top-down approaches to, 39–40
Valencia, Spain, research dataset for, 247
Valle Theatre, Rome, 82
Venice, Italy, research dataset for, 247
Viladecans, Spain, research dataset for, 247
Villa della Piscina (Rome), 230
Village Community Movement (VCM), 220
Villas Ad Duas Lauros (Rome), 230
Villeurbanne, France, research dataset for, 247

Walker, Latrice, 179
Washington, DC
 Advisory Neighborhood Commissions, 165
 land value in, 37
 limited-equity cooperative ownership in, 86–87
 vacancy rates in, 48
Water services management
 Naples resolutions, 135
 Ostrom's findings on, 78
Wealth and resource disparities, 2–3, 212–218. *See also* Inequalities; "Right to the city" concept; Urban poor
 environmental justice concerns and, 35

as new urban crisis, 7–8
rise of neoliberalism and, 158
tech justice, 212–218, 244–254
Wellington, New Zealand, research dataset for, 252
What about the Children, 179
White flight, 222
Wien, Austria, research dataset for, 247
Wikipedia, 202
Wireless mesh networks. *See* Community wireless mesh networks (CWN)
Woelab, 220
Won-Soon, Park, 113
Workers' Party, 168
Wynberg, Rachel, 171

Yogiakarta, Indonesia, research dataset for, 251

Zaragoza, Spain, research dataset for, 247
Zero Waste in Shared Space challenge, 179
Zoning plans, 8, 73, 92, 144, 165

Urban and Industrial Environments
Series editor: Robert Gottlieb, Henry R. Luce Professor of Urban and Environmental Policy, Occidental College

Maureen Smith, *The U.S. Paper Industry and Sustainable Production: An Argument for Restructuring*

Keith Pezzoli, *Human Settlements and Planning for Ecological Sustainability: The Case of Mexico City*

Sarah Hammond Creighton, *Greening the Ivory Tower: Improving the Environmental Track Record of Universities, Colleges, and Other Institutions*

Jan Mazurek, *Making Microchips: Policy, Globalization, and Economic Restructuring in the Semiconductor Industry*

William A. Shutkin, *The Land That Could Be: Environmentalism and Democracy in the Twenty-First Century*

Richard Hofrichter, ed., *Reclaiming the Environmental Debate: The Politics of Health in a Toxic Culture*

Robert Gottlieb, *Environmentalism Unbound: Exploring New Pathways for Change*

Kenneth Geiser, *Materials Matter: Toward a Sustainable Materials Policy*

Thomas D. Beamish, *Silent Spill: The Organization of an Industrial Crisis*

Matthew Gandy, *Concrete and Clay: Reworking Nature in New York City*

David Naguib Pellow, *Garbage Wars: The Struggle for Environmental Justice in Chicago*

Julian Agyeman, Robert D. Bullard, and Bob Evans, eds., *Just Sustainabilities: Development in an Unequal World*

Barbara L. Allen, *Uneasy Alchemy: Citizens and Experts in Louisiana's Chemical Corridor Disputes*

Dara O'Rourke, *Community-Driven Regulation: Balancing Development and the Environment in Vietnam*

Brian K. Obach, *Labor and the Environmental Movement: The Quest for Common Ground*

Peggy F. Barlett and Geoffrey W. Chase, eds., *Sustainability on Campus: Stories and Strategies for Change*

Steve Lerner, *Diamond: A Struggle for Environmental Justice in Louisiana's Chemical Corridor*

Jason Corburn, *Street Science: Community Knowledge and Environmental Health Justice*

Peggy F. Barlett, ed., *Urban Place: Reconnecting with the Natural World*

David Naguib Pellow and Robert J. Brulle, eds., *Power, Justice, and the Environment: A Critical Appraisal of the Environmental Justice Movement*

Eran Ben-Joseph, *The Code of the City: Standards and the Hidden Language of Place Making*

Nancy J. Myers and Carolyn Raffensperger, eds., *Precautionary Tools for Reshaping Environmental Policy*

Kelly Sims Gallagher, *China Shifts Gears: Automakers, Oil, Pollution, and Development*

Kerry H. Whiteside, *Precautionary Politics: Principle and Practice in Confronting Environmental Risk*

Ronald Sandler and Phaedra C. Pezzullo, eds., *Environmental Justice and Environmentalism: The Social Justice Challenge to the Environmental Movement*

Julie Sze, *Noxious New York: The Racial Politics of Urban Health and Environmental Justice*

Robert D. Bullard, ed., *Growing Smarter: Achieving Livable Communities, Environmental Justice, and Regional Equity*

Ann Rappaport and Sarah Hammond Creighton, *Degrees That Matter: Climate Change and the University*

Michael Egan, *Barry Commoner and the Science of Survival: The Remaking of American Environmentalism*

David J. Hess, *Alternative Pathways in Science and Industry: Activism, Innovation, and the Environment in an Era of Globalization*

Peter F. Cannavò, *The Working Landscape: Founding, Preservation, and the Politics of Place*

Paul Stanton Kibel, ed., *Rivertown: Rethinking Urban Rivers*

Kevin P. Gallagher and Lyuba Zarsky, *The Enclave Economy: Foreign Investment and Sustainable Development in Mexico's Silicon Valley*

David N. Pellow, *Resisting Global Toxics: Transnational Movements for Environmental Justice*

Robert Gottlieb, *Reinventing Los Angeles: Nature and Community in the Global City*

David V. Carruthers, ed., *Environmental Justice in Latin America: Problems, Promise, and Practice*

Tom Angotti, *New York for Sale: Community Planning Confronts Global Real Estate*

Paloma Pavel, ed., *Breakthrough Communities: Sustainability and Justice in the Next American Metropolis*

Anastasia Loukaitou-Sideris and Renia Ehrenfeucht, *Sidewalks: Conflict and Negotiation over Public Space*

David J. Hess, *Localist Movements in a Global Economy: Sustainability, Justice, and Urban Development in the United States*

Julian Agyeman and Yelena Ogneva-Himmelberger, eds., *Environmental Justice and Sustainability in the Former Soviet Union*

Jason Corburn, *Toward the Healthy City: People, Places, and the Politics of Urban Planning*

JoAnn Carmin and Julian Agyeman, eds., *Environmental Inequalities Beyond Borders: Local Perspectives on Global Injustices*

Louise Mozingo, *Pastoral Capitalism: A History of Suburban Corporate Landscapes*

Gwen Ottinger and Benjamin Cohen, eds., *Technoscience and Environmental Justice: Expert Cultures in a Grassroots Movement*

Samantha MacBride, *Recycling Reconsidered: The Present Failure and Future Promise of Environmental Action in the United States*

Andrew Karvonen, *Politics of Urban Runoff: Nature, Technology, and the Sustainable City*

Daniel Schneider, *Hybrid Nature: Sewage Treatment and the Contradictions of the Industrial Ecosystem*

Catherine Tumber, *Small, Gritty, and Green: The Promise of America's Smaller Industrial Cities in a Low-Carbon World*

Sam Bass Warner and Andrew H. Whittemore, *American Urban Form: A Representative History*

John Pucher and Ralph Buehler, eds., *City Cycling*

Stephanie Foote and Elizabeth Mazzolini, eds., *Histories of the Dustheap: Waste, Material Cultures, Social Justice*

David J. Hess, *Good Green Jobs in a Global Economy: Making and Keeping New Industries in the United States*

Joseph F. C. DiMento and Clifford Ellis, *Changing Lanes: Visions and Histories of Urban Freeways*

Joanna Robinson, *Contested Water: The Struggle Against Water Privatization in the United States and Canada*

William B. Meyer, *The Environmental Advantages of Cities: Countering Commonsense Antiurbanism*

Rebecca L. Henn and Andrew J. Hoffman, eds., *Constructing Green: The Social Structures of Sustainability*

Peggy F. Barlett and Geoffrey W. Chase, eds., *Sustainability in Higher Education: Stories and Strategies for Transformation*

Isabelle Anguelovski, *Neighborhood as Refuge: Community Reconstruction, Place Remaking, and Environmental Justice in the City*

Kelly Sims Gallagher, *The Globalization of Clean Energy Technology: Lessons from China*

Vinit Mukhija and Anastasia Loukaitou-Sideris, eds., *The Informal American City: Beyond Taco Trucks and Day Labor*

Roxanne Warren, *Rail and the City: Shrinking Our Carbon Footprint While Reimagining Urban Space*

Marianne E. Krasny and Keith G. Tidball, *Civic Ecology: Adaptation and Transformation from the Ground Up*

Erik Swyngedouw, *Liquid Power: Contested Hydro-Modernities in Twentieth-Century Spain*

Ken Geiser, *Chemicals without Harm: Policies for a Sustainable World*

Duncan McLaren and Julian Agyeman, *Sharing Cities: A Case for Truly Smart and Sustainable Cities*

Jessica Smartt Gullion, *Fracking the Neighborhood: Reluctant Activists and Natural Gas Drilling*

Nicholas A. Phelps, *Sequel to Suburbia: Glimpses of America's Post-Suburban Future*

Shannon Elizabeth Bell, *Fighting King Coal: The Challenges to Micromobilization in Central Appalachia*

Theresa Enright, *The Making of Grand Paris: Metropolitan Urbanism in the Twenty-first Century*

Robert Gottlieb and Simon Ng, *Global Cities: Urban Environments in Los Angeles, Hong Kong, and China*

Anna Lora-Wainwright, *Resigned Activism: Living with Pollution in Rural China*

Scott L. Cummings, *Blue and Green: The Drive for Justice at America's Port*

David Bissell, *Transit Life: Cities, Commuting, and the Politics of Everyday Mobilities*

Javiera Barandiarán, *From Empire to Umpire: Science and Environmental Conflict in Neoliberal Chile*

Benjamin Pauli, *Flint Fights Back: Environmental Justice and Democracy in the Flint Water Crisis*

Karen Chapple and Anastasia Loukaitou-Sideris, *Transit-Oriented Displacement or Community Dividends? Understanding the Effects of Smarter Growth on Communities*

Henrik Ernstson and Sverker Sörlin, eds., *Grounding Urban Natures: Histories and Futures of Urban Ecologies*

Katrina Smith Korfmacher, *Bridging the Silos: Collaborating for Environment, Health, and Justice in Urban Communities*

Jill Lindsey Harrison, *From the Inside Out: The Fight for Environmental Justice within Government Agencies*

Anastasia Loukaitou-Sideris, Dana Cuff, Todd Presner, Maite Zubiaurre, and Jonathan Jae-an Crisman, *Urban Humanities: New Practices for Reimagining the City*

Govind Gopakumar, *Installing Automobility: Emerging Politics of Mobility and Streets in Indian Cities*

Amelia Thorpe, *Everyday Ownership: PARK(ing) Day and the Practice of Property*

Tridib Banerjee, *In the Images of Development: City Design in the Global South*

Ralph Buehler and John Pucher, eds., *Cycling for Sustainable Cities*

Casey J. Dawkins, *Just Housing: The Moral Foundations of American Housing Policy*

Kian Goh, *Form and Flow: The Spatial Politics of Urban Resilience and Climate Justice*

Kian Goh, Anastasia Loukaitou-Sideris, and Vinit Mukhija, eds., *Just Urban Design: The Struggle for a Public City*

Sheila R. Foster and Christian Iaione, *Co-Cities: Innovative Transitions toward Just and Self-Sustaining Communities*